RENEWABLE ENERGY: RESEARCH, DEVELOPMENT AND POLICIES

RENEWABLE ENERGY PRODUCTS

LABELING, PRODUCTION GUIDANCE AND CONSUMER ATTITUDES

Renewable Energy: Research, Development and Policies

Additional books in this series can be found on Nova's website under the Series tab.

Additional E-books in this series can be found on Nova's website under the E-book tab.

Energy Science, Engineering and Technology

Additional books in this series can be found on Nova's website under the Series tab.

Additional E-books in this series can be found on Nova's website under the E-book tab.

RENEWABLE ENERGY: RESEARCH, DEVELOPMENT AND POLICIES

RENEWABLE ENERGY PRODUCTS

LABELING, PRODUCTION GUIDANCE AND CONSUMER ATTITUDES

BRETT R. HARRIS
EDITOR

nova publishers
New York

For permission to use material from this book please contact us:
Telephone 631-231-7269; Fax 631-231-8175
Web Site: http://www.novapublishers.com

Library of Congress Cataloging-in-Publication Data

ISBN: 978-1-62417-565-7

Published by Nova Science Publishers, Inc. † New York

CONTENTS

PREFACE

Green marketing – a marketing strategy highlighting the environmental attributes of a product – dates back to the 1970s but began to flourish in the early 1990s. More recently, a number of companies using renewable energy in the manufacture of products have begun to communicate renewable energy use directly on product packaging, relying either on a logo or some combination of text and imagery. This book describes the experience of companies that communicate to consumers that products are "made with renewable energy." Corporate commitments to using renewable energy, and communicating that commitment on product packaging and through other means, could play an important role in educating consumers about the availability and feasibility of using renewable energy as an alternative energy source.

Chapter 1 - Researchers identified nearly 50 companies that communicate renewable energy use on product packaging. Representatives from 20 companies were interviewed and asked to discuss their experiences marketing products produced with renewable energy. Companies take a variety of approaches to communicating to consumers that their products are made with renewable energy. Some companies are labeling on-product, while others are using their websites, social media, and other promotions. On-product labeling can include use of a logo or use of imagery and statements, such as a picture of a wind turbine and note that the product is made with renewable energy.

Why are companies promoting the use of renewable energy on products? The primary motivation identified was to communicate to the consumer about the company's commitment to renewable energy and, in doing so, enhance the image of the brand. Other motivations included differentiating a product, targeting environmentally conscious consumers, and to a lesser degree, following an existing industry trend and earning a price premium.

What challenges do companies face when making on-product claims about renewable energy? One of the primary challenges identified was the limited and competing uses of physical space on a product ("product real estate"). Other challenges included determining the appropriate language and content to include and knowing whether consumers will recognize and understand renewable energy messages. Costs associated with modifying packaging, costs of certification and program requirements for use of a third-party logo, and international product marketing were identified as minor challenges.

The future growth rate of this new market for labeling products with renewable energy claims remains to be seen. While some lessons can be learned from how eco-products have been marketed historically, products that are made with renewable energy are unique in that they are identical to the comparable conventional product in quality and performance; only the energy used to manufacture the product is different.

Chapter 2 - Many innovative businesses have successfully introduced "green" products in recent years. While many of these products have saved money, they have also reduced impacts on human health and the environment by using recycled or recyclable material (e.g., recycled paper products), reducing their energy usage (e.g., efficient washing machines that use less water and energy) or eliminating the use of toxic chemicals during their manufacture, for example. How do these business owners and product designers go about improving existing products or introducing new green products? What motivates them to do so? In providing insight into these questions, this guidance document aims to encourage you to identify opportunities to green your existing products or introduce new green products to your product line.

This guidance document focuses on opportunities to help your business reduce the impact on the environment, while saving you money. Two approaches for improvement are product improvement and process improvement.

The goal for these two approaches is to benefit your bottom line, while also benefiting the environment. They aim to improve your profitability in a variety of ways, as shown in the table below.

It is helpful to consider the product development process (Figure 2) when considering how to incorporate greening concepts. While greening concepts should be incorporated into all product development phases, special consideration should be given to the product's design – studies show that about 70% of a product's environmental impacts are determined by decisions made during the design stage.

Chapter 3 - The data in this report are taken from Natural Marketing Institute's (NMI) Lifestyles of Health and Sustainability (LOHAS) Consumer Trends Database. Created in 2002, the syndicated consumer database contains responses from 2,000 to 4,000 nationally representative US adults (meaning the demographics of the sample are consistent with US Census findings) each year.

A sample of 2,000 has a confidence interval of +/-2.2 percentage points at the 95% confidence level, which decreases to +/- 1.2 percentage points with 4,000 respondents. NMI used the database to analyze consumer attitudes and behavior related to renewable energy and to update previously conducted related research. Specifically, this report will explore consumer awareness, concerns, perceived benefits, knowledge of purchase options, and usage of renewable energy as well as provide regional comparisons and trends over time.

In: Renewable Energy Products
Editor: Brett R. Harris
ISBN: 978-1-62417-565-7

Chapter 1

MADE WITH RENEWABLE ENERGY: HOW AND WHY COMPANIES ARE LABELING CONSUMER PRODUCTS*

Deborah Baker Brannan, Jenny Heeter and Lori Bird

LIST OF ACRONYMS AND ABBREVIATIONS

BD	Becton, Dickinson and Company
CRL	carbon reduction label
DOE	U.S. Department of Energy
EERE	Energy Efficiency and Renewable Energy
EPA	U.S. Environmental Protection Agency
FTC	Federal Trade Commission
GWh	gigawatt-hour
kW	kilowatt
kWh	kilowatt-hour
NREL	National Renewable Energy Laboratory
QR	quick response
REC	renewable energy certificate
USDA	U.S. Department of Agriculture

* This is an edited, reformatted and augmented version of the U.S. Department of Energy, Office of Energy Efficiency & Renewable Energy, NREL/TP-6A20-53764, dated March 2012.

EXECUTIVE SUMMARY

Green marketing—a marketing strategy highlighting the environmental attributes of a product—dates back to the 1970s but began to flourish in the early 1990s. More recently, a number of companies using renewable energy in the manufacture of products have begun to communicate renewable energy use directly on product packaging, relying either on a logo or some combination of text and imagery. This report discusses the experience of companies that communicate to consumers that products are "made with renewable energy." Corporate commitments to using renewable energy, and communicating that commitment on product packaging and through other means, could play an important role in educating consumers about the availability and feasibility of using renewable energy as an alternative energy source.

Researchers identified nearly 50 companies that communicate renewable energy use on product packaging. Representatives from 20 companies were interviewed and asked to discuss their experiences marketing products produced with renewable energy. Companies take a variety of approaches to communicating to consumers that their products are made with renewable energy. Some companies are labeling on-product, while others are using their websites, social media, and other promotions. On-product labeling can include use of a logo or use of imagery and statements, such as a picture of a wind turbine and note that the product is made with renewable energy.

Why are companies promoting the use of renewable energy on products? The primary motivation identified was to communicate to the consumer about the company's commitment to renewable energy and, in doing so, enhance the image of the brand. Other motivations included differentiating a product, targeting environmentally conscious consumers, and to a lesser degree, following an existing industry trend and earning a price premium.

What challenges do companies face when making on-product claims about renewable energy? One of the primary challenges identified was the limited and competing uses of physical space on a product ("product real estate"). Other challenges included determining the appropriate language and content to include and knowing whether consumers will recognize and understand renewable energy messages. Costs associated with modifying packaging, costs of certification and program requirements for use of a third-party logo, and international product marketing were identified as minor challenges.

The future growth rate of this new market for labeling products with renewable energy claims remains to be seen. While some lessons can be learned from how eco-products have been marketed historically, products that are made with renewable energy are unique in that they are identical to the comparable conventional product in quality and performance; only the energy used to manufacture the product is different.

1. INTRODUCTION

Green marketing—a marketing strategy highlighting the environmental attributes of a product, often through the use of labels or logos—dates back to the 1970s. It did not proliferate until the 1990s (Rex and Baumann 2007), however, when extensive market research identified a rapidly growing group of consumers with a heightened concern for the environment. Consumers expressed not only a preference for green products but also a willingness to pay a premium for such products (Peattie and Crane 2005). The response was a surge in green marketing that lasted through the early 1990s. By the mid-1990s, however, the green marketing rush had waned. The market for green products had remained relatively weak due to the apparent gap between environmental concern and sales suggested by consumer surveys to actual sales achieved (Rex and Baumann 2007; Peattie and Crane 2005).

Although green products remain a niche market, more than 80% of consumers continue to express interest in protecting the environment, even though those levels have declined slightly in recent years (Natural Marketing Institute 2011). As such, companies continue to pursue new ways of communicating their actions to protect the environment to consumers. Providing information on the type of energy used to produce a product—specifically whether it is made with renewable energy—is an emerging strategy. Survey data suggest that consumers view renewable energy favorably and prefer products made with renewable energy sources. Surveys conducted by the Natural Marketing Institute, for example, found that 80% of consumers in 2010 indicated that they care about the use of renewable energy (Natural Marketing Institute 2011). Additionally, a recent poll conducted by Vestas and TNS Gallup found that 65% of consumers worldwide prefer products that are made with wind energy (Vestas and TNS Gallup 2011).

A number of companies using renewable energy in the manufacture of products communicate this directly on product packaging, relying either on a logo or on some combination of text and imagery (e.g., wind turbines). Other companies have refrained from using on-product messaging and rely on other types of marketing collateral to communicate their use of renewable energy, typically via their websites. In general, communicating the renewable energy content of products differs from some earlier green marketing efforts that often focused on modifications made to the products themselves. One reason the adoption of green products has been limited is because green products often do not compete with comparable conventional products on important dimensions, such as price, quality, or performance (Gallastegui 2002). Products that use less material or recycled material, for example, could be viewed as inferior or less effective. In contrast, the use of renewable energy in the manufacture of a product has no impact on the quality of the product itself. Therefore, previous experience with green marketing might not be entirely transferable to this new strategy of communicating the renewable energy used in manufacturing a product.

This report discusses the experience of companies that communicate to consumers that products are "made with renewable energy." Representatives from 20 companies were interviewed and asked to discuss their experiences marketing products produced using renewable energy. Interview participants were asked to discuss motivations for making on-product claims and to describe their primary challenges in labeling products with renewable energy use. The small number of interviews would not have provided robust quantitative data; therefore, interviews focused on gathering qualitative responses. The first half of this report provides an overview of the type of companies that have labeled products or advertised them as being made with renewable energy. It also highlights the avenues companies use to describe their use of renewable energy. What follows is a discussion of the motivations for making on-product claims about the use of renewable energy and the challenges in doing so, based on insights learned through the interview process.

2. Experience Offering Products Made with Renewable Energy

The environmental attributes of a product increasingly factor into the consumer's purchasing decision—including the type of energy used to manufacture the product. A recent global consumer survey revealed that 65% of consumers worldwide would prefer a product made with wind energy (Vestas and TNS Gallup 2011). Companies have begun offering products that are made with renewable energy. Highlighting the energy source used to make the product allows a company to differentiate its product, communicate its environmental activities and sustainability ethic, and enhance the image of the brand. For the renewable energy industry, corporate interest in promoting the use of renewable energy in the manufacture of products helps educate consumers about the benefits of renewable energy and demonstrate that it is an option for electricity generation.

Although there is no comprehensive database identifying all companies in the United States that advertise that their products are made with renewable energy, researchers identified nearly 50 companies that do so. Thirty-eight companies are members of the Green-e Marketplace[1] and at least nine companies make independent claims of using renewable energy: Aveda, Boulder Canyon Chips, Drew's (salad dressing), Kettle Brand (potato chips), Lundberg Family Farms, New Belgium Brewing, Kraft, SC Johnson, and WhiteWave Foods. Appendix A provides a sample of these companies, which range in size from small companies to large multinational corporations and cover a wide spectrum of industries. Some companies are in the business-to-business sector and others are in the business-to-consumer sector.

The extent to which companies market the use of renewable energy in the manufacturing of products varies considerably. Some companies use renewable energy and make claims on one product, and others label a single line or multiple product lines. Of SC Johnson's Ziploc product line, for example, only the Ziploc Evolve carries a statement that the product is "made with wind energy."

Other companies advertise their use of renewable energy more broadly and apply it to an entire product line. Cascades Tissue Group's North River brand, for example—which includes toilet paper, hand towels, and other tissue products—all carry the Green-e logo. The oldest and most well-known third-party certified renewable energy logo in the United States is the Green-e logo. Companies using 100% renewable energy for their manufacturing facilities, as well as a minimum amount for their entire company, are eligible to use the Green-e Marketplace logo, though not all companies choose to do so.[2] Fifty-one organizations participated in the Green-e Marketplace program at some point in 2010, with 500 products achieving certification and carrying the Green-e logo (Center for Resource Solutions 2010).

For more information on Green-e, refer to Text Box 1. Additionally, some companies use labels or other means to advertise that all of their products are made using renewable energy. For example, all Intelligent Nutrients health and beauty products carry the Green-e logo.

Companies in both business-to-business and business-to-consumer sectors have communicated their use of renewable energy on specific product lines or through a broader corporate branding initiative.

Companies in the business-to-business sector have a tendency to use renewable energy labeling for a specific product or product line. For example, Cascades Tissue Group advertises its use of renewable energy as part of its marketing strategy for the North River brand of tissue products, which is sold exclusively to commercial and industrial customers. Some companies, however, are taking a corporate branding approach. For example, BD (Becton, Dickinson and Company, a medical technology company) has incorporated the use of logos to convey its use of renewable energy into its corporate brand and anticipates extensive use of a renewable energy product logo in the years ahead.

The majority of products made with renewable energy are designed to appeal to the environmentally conscious consumer, such that being "made with renewable energy" simply adds to the image of the product, product line, or corporate brand. The Ziploc Evolve plastic bag, for example, uses 25% less plastic than comparable products, in addition to being manufactured partially using wind energy. Among food products, "natural" food products are more likely to be labeled as being made with renewable energy. WhiteWave Foods, for example, advertises its involvement with renewable energy on Silk and Horizon Organics products but does not include on-product messaging on its International Delight products—which are not "natural" food products—even though International Delight products also are made using renewable energy. The International Delight website does include information about renewable energy use.

Some products and companies have neither a natural connection to the environmental consumer or the ability to easily integrate onsite renewable generation facilities. For example, BD purchases 100% of its electricity for one of its manufacturing facilities through a utility green power program and expects to label products, such as hypodermic needles, in the near future.

Some companies do not sell any products and instead provide services, including services that do not have a natural connection to the environmental consumer. HostPapa offers Internet hosting services, for example, and iMoveGreen and iStoreGreen provide moving and storage services. These companies tend to use renewable energy to brand themselves as being committed to renewable energy and environmental sustainability and, as a result, appeal to the environmentally conscious consumer.

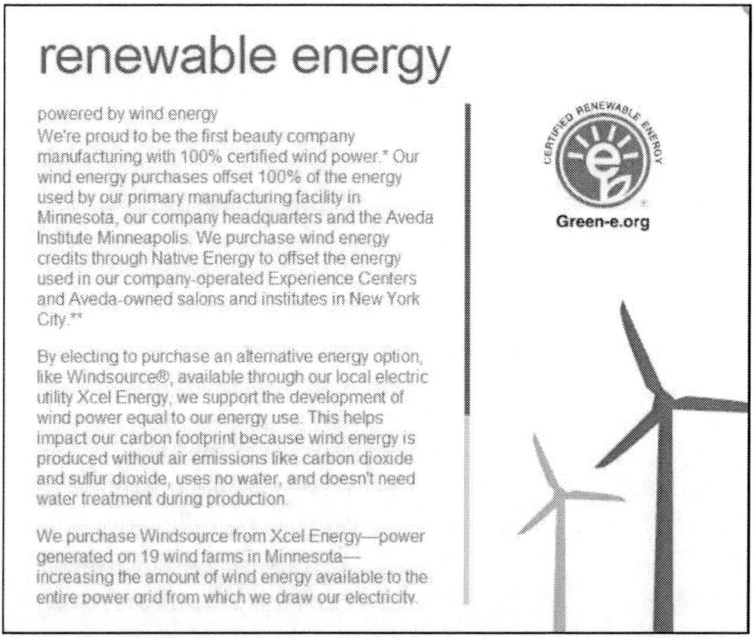

Source: Aveda Corporation

Figure 1. Aveda's website content on renewable energy.

3. Communicating the Use of Renewable Energy

Companies take a variety of approaches for communicating to consumers that their products are made with renewable energy. The traditional method to highlight the use of renewable energy in the manufacturing of a product is to show the information directly on the product or product packaging. "Packaging is a real and tangible way to engage consumers," said Julie Berling, director of brand advocacy for GNP Company. GNP Company sells premium chicken products and, according to Bering, on-product messaging is important because "It's really the first introduction consumers have with our brand, and if they like what they see, they're more likely to try it" (Berling2011).

On-product messaging has its challenges, however. The limited real estate available on product packaging can make it difficult to incorporate renewable energy logos, imagery, or information on the package. Aside from logistical complications, on-product messaging faces other challenges. For one, on-product messaging might not be effective. "People are busy, and may not be paying attention" (Berling2011). Consumers also might be hesitant to trust on-product messaging due to a growing concern over "green washing." For these reasons, some companies have avoided on-product messaging altogether. The beauty product company Aveda, for example, uses some on-product messaging but also advertises its commitment to renewable energy through outlets including its website (Figure 1), Facebook, salons, and stores.

3.1. On-Product Messaging

There are two classes of on-product messaging: (1) imagery and statements and (2) logos. In the first category, companies have opted to use imagery (primarily of wind turbines), which generally is accompanied by a statement describing the company's involvement with renewable energy. Strathmore Artists Windpower Watercolor paper features a wind turbine on the cover of the package as well as a description of the company's renewable energy use on the inside cover (Figure 2).

Source: Strathmore Artists.

Figure 2. Strathmore Artists Windpower Watercolor paper.

Companies vary in the amount of information that accompanies imagery on products. Some companies include a one-sentence statement and others include entire paragraphs. Kettle Brand, for example, uses the statement "And they're backed by a commitment to sustainability like wind power, solar power, green building, and biodiesel," which is accompanied by an image of the sun. Conversely, WhiteWave Food's Silk soymilk uses most of the side panel on the carton and includes images of wind turbines and a paragraph describing the company's involvement with renewable energy.

Rather than using imagery and statements, many companies usea renewable energy logo, similar to the USDA Organic logo or the Dolphin-Safe logo. A label or logo that is recognizable, easy to understand, and trusted is more effective at conveying information to the consumer and, as a result, contributes to market growth (Rex and Baumann 2007). Companies that rely on a logo use either an independent logo (one that is self-designed) or a third-party verified logo. Each approach has advantages and disadvantages.

A common concern when using any type of logo is that consumers might not recognize the logo or might not understand what it means. By using an independent logo, a company can develop something it thinks will effectively engage its consumers. Independent logos typically make use of strong imagery, incorporating a wind turbine, a green leaf, or the planet Earth. These logos generally include text such as "made with renewable energy" or "made with green energy." A company might be inclined to use its own logo instead of paying a licensing fee to use a third-party logo. Although the vast majority of interviewed companies that use a third-party logo did not identify licensing costs as significant, a few companies did comment that because it was difficult to quantitatively identify the benefits associated with using the logo, it was a line-item in the budget that could come under scrutiny.

Source: Center for Resource Solutions.

Figure 3. Santa Cruz Organics Green-e labeling.

Third-party logos could be beneficial in that they can provide credibility to the renewable energy claim. Credibility is highly desirable by companies, as they seek to avoid potential negative publicity associated with "green washing." Laura M. Thompson, Ph.D, Director of Sustainability and Technical Marketing for Sappi Fine Paper North America, commented that third-party certifications offer additional assurances to customers. Thompson says, "At Sappi, we routinely make claims about our high levels of renewable energy usage but many of our customers also want a third-party logo attesting to this certification"(Thompson 2011). Among other companies, Santa Cruz Organics and Cascades Tissue Group both use the third-party Green-e logo on their products (see Figure 3 and Figure 4).

Source: Cascades Tissue Group.

Figure 4. Cascades Tissue Group's Green-e labeling.

Text Box 1. Green-e Marketplace Program

The Green-e program was developed in 1998 by the Center for Resource Solutions to provide transparency and accountability to the voluntary renewable energy market. Green-e Energy and Green-e Climate verify and certify renewable energy and carbon offsets, respectively. As more and more businesses turned to Green-e for third-party certification of on-site generation or the purchase of third-party certified RECs, the Center for Resource Solutions developed the Green-e Marketplace in 2004. This provides businesses with a way of communicating their commitment to renewable energy to their customers. Businesses meeting a minimum renewable energy requirement for their U.S. operations using Green-e Energy certified renewable energy may display the Green-e logo on annual reports, sustainability reports, and websites.

To be permitted to display the logo directly on a product, product packaging, and product marketing materials, a company must meet the additional requirement that the product must have been manufactured using 100% renewable energy.

Since the inception of the Green-e Marketplace, the number of participating businesses and the number of products carrying the logo both have grown. In 2004, 40 companies were members of Green-e Marketplace and only 12 products carried the Green-e logo (Center for Resource Solutions 2004). By 2010, these numbers had risen to 51 companies and more than 500 products (Center for Resource Solutions 2010). These companies alone accounted for approximately 394 gigawatt-hours (GWh) of renewable electricity generation in 2010 (Center for Resource Solutions2011).

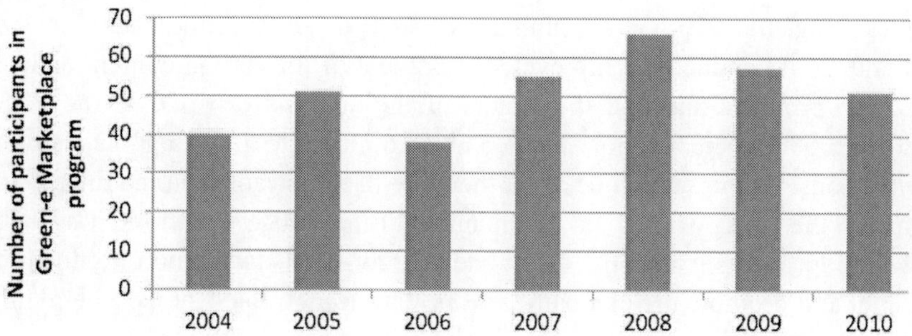

Figure 5. Number of participants in the Green-e Markeplace Program, 2004–2010.

Members of Green-e Marketplace span a range of industries, including printing and packaging, food and beverage, health and beauty, and Internet data centers. In addition to the wide variety of industries making "made with renewable energy" claims, the members of Green-e Marketplace include both small and large companies. Companies such as Office Depot, SC Johnson, Intel, and Pepsi all have been members of Green-e Marketplace.

In some cases, rather than using the Green-e logo, companies have opted to use the logos of their Green-e Energy certified REC providers. GNP Company's Just BARE Chicken product line formerly carried the Renewable Choice Energy logo.

Text Box 2. GNP Company

GNP Company is a producer of premium chicken products in the Midwest. The company was founded in 1926 and has been owned and operated by three generations of the Helgeson family. It has a strong commitment to family farming and environmental sustainability. GNP Company has two premium, natural brands: Gold'n Plump and Just BARE. The Just BARE product line was launched in 2008 in response to the consumer demand for fresh chicken products made to meet higher standards of natural products that research identified. The Just BARE product line is intended to meet this consumer demand by offering a premium quality product with a commitment to environmental sustainability. In particular, the Just BARE brand only uses chickens raised on vegetarian feed and no antibiotics and strives for more sustainable packaging. In 2010, GNP Company wanted to reinforce the claims of the brand, so it conducted a life cycle assessment of the carbon content of Just BARE products and had the results of the analysis certified to the PAS 2050:2008 by the Carbon Trust and committed to reducing the carbon emissions. At the same time, it partnered with Renewable Choice Energy to match the energy used in the manufacturing process with RECs. Just BARE products currently carry the Carbon Trust's Carbon Reduction Label and promote the company's work with Renewable Choice Energy on its website and in various marketing materials.

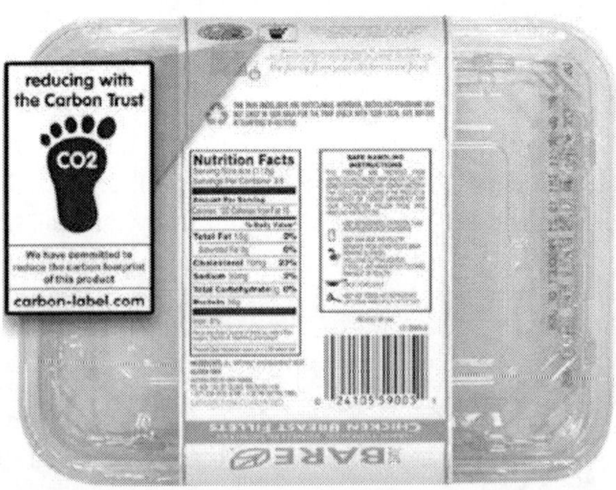

Figure 6. Carbon Trust's Carbon Reduction Label on Just BARE's packaging.

In addition to the on-product messaging, the Just BARE brand actively promoted its involvement with renewable energy through other means. A promotion called "Learn to Earn," for example, enabled consumers to answer questions on sustainability on the company's website. If a customer answered one question correctly, they earned a coupon. The company used on-pack stickers and Facebook to promote the "Learn to Earn" campaign, and also has used blogs in its effort to engage and educate consumers.

Just BARE cited both product differentiation and brand enhancement as the primary motivations for using the renewable energy logo. The biggest challenge associated with on-product labeling is the limited and competing uses of product packaging real estate. Julie Berling commented that having on-product messaging is critical, "If we can't show it on the package, it won't motivate a purchase." There are many competing uses of product packaging, however, and not just for marketing. As a producer of fresh chicken products, GNP Company's brands are required to include safe handling instructions on the package.

As an emerging marketing tool, new logos are continuously entering the marketplace. Recently, the non-profit organization WindMade published the first global renewable energy label for companies using wind power in their operations. Windmade's lead sponsor and founding partner is Vestas.[3] Businesses that want to use the WindMade label must procure a minimum of 25% of their power demand from wind power. Procurements are verified by a third-party. Businesses can use two different types of labels, one that focuses on the wind power used and one that mentions the use of other renewable technologies (Figure 7). WindMade is currently developing a standard for labeling products, which is scheduled to be launched in the summer of 2012.

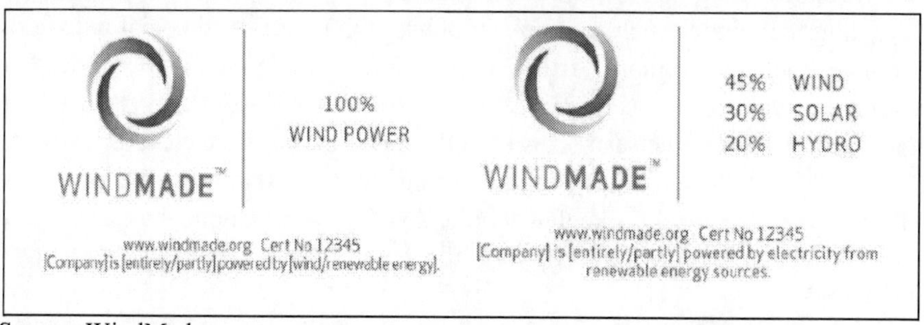

Source: WindMade.

Figure 7. WindMade's sample logos for business use.

Some of the companies interviewed highlighted that a multi-attribute label could address concerns regarding limited packing real estate and could simplify the certification process on the company end. A multi-attribute label could address multiple environmental attributes—not just the use of renewable energy. There has been some preliminary discussion of multi-attribute eco-labeling; however, many challenges would face such a development, including how the standards would be set, which attributes to include, and how the FTC would treat such a label.

3.2. Websites, Social Media, and Other Promotions

Regardless of whether a company uses on-product messaging, it typically relies on its website as an additional means of communication. Websites have minimal constraints, so companies may include detailed descriptions of their involvement with renewable energy. In conventional marketing efforts, product labels and logos are just one technique in the marketers' toolbox (Rex and Baumann 2007). To effectively market a product and spur market growth, a variety of marketing strategies should be adopted (e.g., advertisements, publicity, and sales promotions).[4] Companies often describe their onsite generation facilities, green power purchases, and REC purchases through these means. Interestingly, some service-based companies also communicate their use of renewable energy through websites. Such companies do not have a product per se, however, and rely exclusively on their websites and other venues to communicate their involvement with renewable energy. Internet-hosting company HostPapa, for example, has branded itself as a "green" hosting company.

In addition to websites, some companies are using more modern marketing techniques such as quick response (QR) codes and social networks like Facebook. Strathmore Artists Paper, for example, uses QR codes so consumers wanting to learn more about how renewable energy is used to manufacture the product can scan the QR code and be directed to a website. The rise of social media has prompted companies to also advertise products through social networks like Facebook. Aveda, for example, has created a Facebook page devoted specifically to the company's renewable energy purchases.

Companies also are using promotions specifically focused on renewable energy as a way to engage and educate consumers. The GNP Company, for example, launched a "Learn to Earn" promotion (see Text Box 2) in early

2011. Since the launch, it has been one of the most visited sections of the company's website. Whitewave, owner of the Silk, Horizon, and International Delight brands of dairy products, developed the "Green Caps" program, in which Whitewave makes a financial donation to the Bonneville Environmental Foundation or Farm Aid for each consumer who purchases a Silk product with a green cap and registers the cap online. Deanna Bratter of Whitewave commented on the effectiveness of promotions in reaching new consumers:

> We have different types of consumers for our Silk products. The consumers we are currently trying to engage are people who are interested in natural and organic but have never tried soymilk, or they aren't really sure of the benefits of organic. A promotion specifically targeted around green energy is a good way to educate them and help drive their purchase decisions (Bratter2011).

Approximately 300,000 green caps were donated through the Green Caps program to either Bonneville Environmental Foundation or Farm Aid.

Companies using renewable energy are increasingly communicating with consumers in a variety of ways so that consumers can access the information when they want to and in the manner they prefer. "It is critically important to have a multi-pronged approach because people consume media in very different ways" (Berling2011). Much uncertainty remains, however, regarding whether consumers will be receptive to this type of strategy and how far-reaching it could be. At this point, it is unclear whether products made with renewable energy will appeal only to a niche consumer segment or instead have a broader appeal.

3.3. Carbon Labels for Consumer Products

Carbon labels are increasingly being used to reflect a product's carbon emissions, which can include a renewable energy component. Although broader than renewable energy labels, constituting a different type of claim, companies using carbon labels may use renewable energy to reduce the carbon footprint of products. The past decade has witnessed a growing interest in carbon product labeling worldwide and examples can be found in the United Kingdom, the United States, Canada, Switzerland, Japan, South Korea, and Thailand, among others.

Although renewable energy labeling often can be done in-house and potentially can be verified by a third-party, labeling the carbon content of products depends on third-party systems to account for the carbon content. Perhaps the most recognized carbon label is the Carbon Trust's carbon reduction label (CRL) or the "black footprint" label, prominent in the United Kingdom and other European markets and just emerging in U.S. markets. As more businesses began to measure and reduce their emissions, the Carbon Trust developed a product label for companies to use to effectively communicate their actions. To use the label, companies must undergo a life cycle analysis to determine the carbon footprint of the product and must make a commitment to reduce their carbon emissions reduction. The "black footprint" takes on two forms, a label or a logo. The CRL indicates that the product lifecycle GHG emissions have been calculated and undergone third-party certification, and the parent company is committed to reducing the life cycle GHG emissions of the product. Inclusion of the product carbon footprint number is optional. Figure 8 shows the CRL label used by Tesco on its orange juice products.

Source: Carbon-label.com.

Figure 8. Tesco's Carbon Trust labeling.

Source: Carbonfund.org.

Figure 9. Carbonfund.org's Carbonfree label.

To date, the Carbon Trust has certified approximately 27,000 product carbon footprints worldwide. In the United Kingdom alone, the carbon footprint reduction label appeared on products with a retail value of approximately £2 billion ($3.2 billion) in 2010 (Carbon Trust 2010).

The CRL primarily is a content-based label, but it includes a performance aspect. Namely, companies using the label make a commitment to reduce the life cycle carbon emissions of their product, and only those companies that achieve carbon reductions may continue to use the label. Renewable energy is one way of reducing carbon emissions and meeting reduction goals. The Carbon Trust has worked with approximately 25 companies in the United States, and one-fifth of those companies have considered renewable energy as a means of reducing carbon emissions.

Another type of carbon label focuses on a product achieving carbon neutrality. The Carbonfund.org Carbonfree label can be used on products that achieve carbon neutrality. The life-cycle carbon emissions of the product are measured and then offset using carbon offsets (Figure 9). In the United States, Carbonfund.org is a third-party certifier of carbon offsets, deriving offsets from a combination of renewable energy, energy efficiency, and forestry projects. Current renewable energy projects include wind projects in India and China, a small hydro project in India, and dairy methane and landfill methane projects in the United States.

In 2007, Carbonfund.org began a labeling program for companies that receive a lifecycle assessment, agree to reduce carbon emissions, and offset the carbon footprint of the product. Although the majority of companies allow Carbonfund.org to choose which projects to invest in, of the companies that specify projects, 40% choose renewable energy projects. The Carbonfund.org carbon-neutral label states that the product is "certified carbon free" and companies have the option of including the amount of carbon that was offset. Currently, about 100 products carry this logo, including products made by Motorola, Domino Sugar, and LG Electronics.

4. Motivations for Promoting Products as Made with Renewable Energy

There are several reasons why a company would choose to invest in or purchase renewable energy. These include reducing or stabilizing energy costs, complying with existing environmental regulations (or as a risk-mitigation strategy against future regulation), and building strong relationships with various stakeholders (Hanson 2005). But why do these companies promote their use of renewable energy on products? Based on interviews with 20 companies, this section of the report identifies some of the motivations. Interview participants were asked to rank each motivation on a five-point scale and comment on any additional motivation not listed; due to the small number of interview participants, results are presented in qualitative form.

4.1. Communicate to the Consumer and Enhance the Image of the Brand

The primary motivation for developing products made with renewable energy is to directly and effectively communicate to the consumer about the company's commitment to renewable energy and, in doing so, enhance the image of the brand. This motivation was particularly important for companies in the business-to-consumer sector, which use renewable energy marketing because it aligns with the corporate brand or product-line image. Great River Organic Milling, for example, commented that its company always has been committed to sustainability and that renewable energy marketing was just an extension of its philosophy (see Text Box 3). Chuck Bennett of Aveda stated that the intention is "not to enhance sales specifically but rather to reinforce the overall environmental commitment of the brand"(Bennett 2011).

4.2. Product Differentiation

Product differentiation proved to be another important motivation for marketing products as "made with renewable energy." Highlighting the energy source used to manufacture the product allows for a point of differentiation. This was particularly important for companies in the business-to-consumer sector that placed claims on one product. As noted previously, these products typically appeal to the environmentally conscious audience, so being made with renewable energy is an additional dimension differentiating the product.

Product differentiation also was cited as an important motivation for some companies in the business-to-business sector. It is important to note that most companies in this situation do not use on-product messaging. Sales are finalized before the customer even receives the product, and it might not make sense to include an on-product message. Many large purchasers, such as Intel, do not use on-product messaging because the product is not an "external product."These companies tend to market their commitment to renewable energy through avenues such as marketing collateral materials, presentations, and events. Interestingly, one business-to-business company we interviewed did use an on-product message. Cascades Tissue Group sells towel and tissue products to the commercial marketplace and includes the Green-e logo because it reminds customers of the environmental attributes of the product.

Text Box 3. Great River Organic Milling

Great River Organic Milling is a small company with just seven employees. The company always has been committed to environmental sustainability and, in 2009, began purchasing RECs to enhance the image of its brand and confirm its corporate philosophy. Rick Halverson, owner of Great River Organic Milling, commented that the ability to purchase RECs was critical in the ability to support the development of renewable energy, "Had REC programs not been available it would have been impossible to justify on-site renewable generation given the size of the company" (Halverson 2011).

Great River Organic Milling became a Green-e Marketplace member in 2009, and the RECs it purchases are Green-e Energy certified. Membership in the Marketplace program enabled Great River Organic Milling to begin using the Green-e logo on all of its products. Great River Organic Milling certainly falls into the category of companies using renewable energy marketing as a corporate branding strategy and, in fact, identified brand enhancement as the primary motivation for using the on-product logo. The biggest challenges associated with using the on-product logo faced by Great River Organic Milling are the costs associated with revising packaging and satisfying the guidelines for using the Green-e logo.

4.3. Targeting Environmentally Conscious Consumers

When asked whether the motivation for marketing products as "made with renewable energy" was to target the environmentally conscious consumer, companies' responses were quite mixed. Some companies stated this was a strong motivation and others said it was not. Companies using renewable energy to market a particular product were more likely to respond that they were trying to target the "green" consumer. Companies that use renewable energy as part of a branding strategy, however, alluded to the fact that their brands already target the "green" consumer; therefore, using a renewable energy marketing message does not attract consumers and instead simply confirms a company's commitment to sustainability.

4.4. Following an Existing Industry Trend

Most companies interviewed stated that they were not motivated by an existing industry trend, instead stating that they were the trendsetters. Given the relatively limited amount of renewable energy marketing, it does seem to be the case that most companies doing this actually are trendsetters, at least in narrowly defined markets. For example, the GNP Company's Just BARE product line is the first fresh chicken product line to carry a "made with renewable energy" label; however, there is experience with communicating the use of renewable energy in the food industry more broadly.

It is important to note that there does seem to be a trend in some of these broadly defined industries. Specifically, several companies in both the paper industry and the food industry have used a renewable energy marketing strategy. In the paper industry, advertising the renewable energy content of products has been adopted by companies in the business-to-business sector selling fine paper (e.g., Sappi Paper and New Leaf Paper) or tissue and toilet paper (e.g., Cascades Tissue Group), as well as by companies in the business-to-consumer sector selling greeting cards (e.g., Marian Heath Greeting Cards) and art paper (e.g., Strathmore Artists Paper). In the food industry, a wide variety of products convey their use of renewable energy, such as chips (e.g., Boulder Canyon Chips), milk (e.g., Whitewave's Silk), cream cheese (e.g., Kraft's Philadelphia Cream Cheese), juice (e.g., Santa Cruz Organics), coffee (e.g., Buywell Coffee), flour (e.g., Great River Organic Milling), rice (e.g., Carolina Plantation Rice), chicken (e.g., GNP Company's Just BARE), and even fresh pasta (e.g., Valley Fine Foods' Pasta Prima).

4.5. Price Premium

A potential motivation for advertising the use of renewable energy in the manufacture of products is to earn a price premium. Research on price premiums reveals an average premium of 10% for sustainable products (Trudel 2011). Furthermore, a recent global consumer survey revealed that

50% of consumers worldwide are willing to pay a premium for products made with renewable energy (Vestas and TNS Gallup 2011). All of the companies interviewed, however, indicated that products marketed as "made with renewable energy" did not earn a price premium. In the tissue industry, Steve Ott of Cascade Tissue Group stated, "It is such a competitive market that there is no real opportunity to earn a price premium" (Ott 2011).

Some products made with renewable energy are premium-quality products and, as a result, earn a price premium. The fine paper industry, for example, produces a higher quality product that costs more to produce and thus earns a price premium. Renewable energy use is one technique for further differentiating these types of products.

Even if the price of products made with renewable energy remains the same, it is possible that companies are able to achieve a reduction in energy costs as a result of investing in onsite renewable energy projects or purchasing through a long-term, fixed, renewable-electricity contract. As a result, their profit margins could increase. Sappi Paper commented that its profit margins did increase for one grade of paper, even though the motivation for developing the onsite renewable energy project and associated paper product was to consistently serve customers and instill customer loyalty rather than increase profit margins (Thompson 2011).

5. Challenges with Product Labeling and Claims

Although purchasing renewable energy to make on-product claims is a relatively new practice, early adopters have learned some lessons. One reason that some companies have chosen to suspend making renewable energy claims on products is because they find the effort is ineffective for the target consumer audience or that a return on investment cannot be documented. Some companies indicated that they are considering suspending their renewable energy labeling efforts due to sagging consumer demand and the inability to quantify the effect of the marketing strategy, particularly in the face of difficult economic conditions. This section explores the challenges associated with communicating the use of renewable energy on products.

Numerous challenges were identified through the interviews conducted for the present research. Similarly to the methodology for assessing motivations, interview participants were asked to rank each challenge on a five-point scale and comment on any additional motivation not listed; due to the small number of interview participants, results are presented in qualitative form. Underlying many conversations was the fundamental challenge of the uncertain response to renewable energy labeling. As with most marketing strategies, it is extremely difficult—if not impossible—to quantitatively identify whether marketing products as "made with renewable energy" is effective at increasing sales, primarily because additional product changes typically are made concurrently.[5] The inability to determine the return on the investment suggests that companies claiming that they use renewable energy on products either are fully committed to it for reasons that extend beyond the potential boost in sales (e.g., corporate image) or are experimenting with the strategy and, depending on sales, ultimately could keep or abandon it.

5.1. Limited and Competing Uses of Product Real Estate

One of the primary challenges associated with making an on-product renewable energy marketing claim is the limited and competing uses of product real estate. Product packaging is carefully designed by marketing professionals to communicate a specific message about the product and the company in the space available. Including too many competing messages can inundate and confuse the consumer, potentially diluting the brand and affecting sales. As a result, companies are careful in choosing which messages to include so that each message resonates with the consumer in a positive way.

Intelligent Nutrients, for example, sells health and beauty products and has products as small as lip balm. Other products have more product real estate, such as Strathmore Artists Paper sketch pads. Strathmore devotes the inside of the sketch pad cover to a conversation on renewable energy (see Figure2).

Aside from the physical constraint of product packaging, the real estate available to describe renewable energy use can be limited by including messages that are required by law. These legal requirements vary by industry. For example, food and household products are required to provide nutritional information, but health and beauty products do not have such requirements.

SC Johnson, for example, is required to include warning messages on its household products, leaving little room for a renewable energy logo.

Given the limited space, marketers must choose which message to include, ensuring that it resonates with the consumer. "The number one issue from a marketing perspective is consumer relevance. Is it meaningful to the consumer?" (Lawson 2011). Although many companies think that an on-product renewable energy message will resonate with their target audience, others focus on alternative messages more relevant to their consumers.

Whitewave's Horizon Organic milk product line, for example, currently only includes renewable energy on-product messaging on its single-serve products. Whitewave states, "It's more a matter of consumer focus at the moment. Currently the package focuses more on the family farmer story and the 'know where your food comes from' movement. It was a marketing decision and not an intentional decision to leave wind out" (Bratter 2011).

5.2. Language and Content

Companies that choose not to use a third-party logo face a greater challenge in determining what content to include when making on-product claims about the use of renewable energy. These companies tend to use some combination of imagery and text. It can be challenging to determine what content to include that will resonate with the consumer in a positive way.

Many companies interviewed commented that imagery is an important component of their marketing strategies. "Walking down the aisle, you have a split second to catch the consumer's attention. [Imagery is] the best way to grab them and tell the story quickly" (Bozek 2011). The imagery used most often in renewable energy marketing claims includes wind turbines, the sun, the planet Earth, and a leaf.

Determining what language to use when including text can be considerably more challenging than selecting imagery, particularly if a company is using RECs. The language surrounding RECs has evolved over time and marketers should take care to change marketing messages as the language evolves to reduce the risk of deceiving consumers. See Text Box 4 for a discussion of how New Belgium Brewing has addressed these issues. Generally, companies have expressed that it can be challenging to determine what language will be relevant to consumers, as "a lot of people just don't get what it means to be offset by renewable energy. There are definitely a lot of challenges in labeling and writing it in a way where it becomes relevant to someone" (Bratter 2011).

Updates to the Federal Trade Commission's (FTC's) *Guide for the Use of Environmental Marketing Claims,* commonly known as the "Green Guides," address renewable energy and carbon marketing claims for the first time and should provide some clarification on the type of language to be used when making a claim about the use of renewable energy. These guides originally were developed in 1992 to address concerns regarding corporate "green washing," or making misleading environmental claims about products. The final Green Guides revisions are expected to be released in 2012, and the degree to which the final guides will differ from the proposed guidance is unclear.

The Green Guides address a variety of issues related to labeling products as being made with renewable energy. The proposed guidance first addresses the type and partial use of renewable energy. It states that marketers should specify the mix and source of renewable energy and, if renewable energy was only used in part of the manufacturing process, clarify which components the claim applies to.

Text Box 4. New Belgium Brewing

In the late 1990s, New Belgium Brewing signed a 10-year contract with its municipal utility to pay a premium for 100% of kilowatt-hours consumed. The commitment helped to support the utility's purchase of wind-derived electricity from Wyoming. Thrilled to have connected to electricity from wind, New Belgium marketed itself as "100% Wind-Powered." As demand for renewable power increased, however, the municipal utility started purchasing wind and non-wind RECs to supplement the wind power. A member of the community (and ex-employee of New Belgium Brewing) brought this to the attention of the media, accusing New Belgium Brewery of green washing.

Katie Wallace of New Belgium comments, "It was never our intention to mislead but we had not kept up with the language as the portfolio changed (Wallace 2011)." New Belgium changed the language on its packaging, eliminating the percentage amount and marketing the product as "Wind Powered, Employee Owned." As New Belgium began investing in its own on-site generation facilities, including solar photovoltaics and bio-gas, it dropped the term "wind" altogether. New Belgium's current marketing message is "Alternatively Empowered" reflecting both the fact that it relies on alternative energy sources and that the company is employee-owned and operated.

The proposed guidance also calls for companies to clarify whether the renewable energy claim applies to the product, the packaging, or particular components of the product or packaging. For example, if a company prints an unqualified renewable energy claim on the report (e.g., "made with renewable energy"), it might be unclear whether it is the company that is using renewable energy or the paper manufacturer. Refer to Text Box 5 for information on how

Green-e is addressing this issue. Due to this potential confusion, the proposed revisions state that unqualified renewable energy claims, such as "made with renewable energy" should be used only when "all, or virtually all, of the significant manufacturing processes used to make the product" rely on renewable generation or RECs, otherwise the claim must be qualified (FTC 2010).

The FTC proposed guidance addresses whether companies can make renewable energy claims when using RECs. Specifically, companies can make renewable energy claims when using RECs and the claim does not have to qualify that RECs are underlying the claim. According to the FTC, "there is no reason to believe that this fact would be material to the consumer" (FTC 2010). The FTC's proposed guidance further specifies that only the owners of the RECs should make renewable energy claims. This is important for companies that have an onsite renewable energy facility but sell the RECs associated with the generation of the system to another party, as they no longer can make renewable energy claims about the system.

Finally, the FTC proposed guidance cautions against the use of unqualified certifications or seals. For example, companies creating their own seal should use "clear and prominent qualifying language to alert the consumers that it created the certifying program" (FTC 2010).

The FTC cites a number of examples of how seals have been used deceptively, including one advertiser's creation and use of a "Consumer Protective Institute" seal. The seal created the false impression that an independent and disinterested organization had approved the product.

5.3. Consumer Recognition and Understanding

Another challenge is whether consumers will recognize and understand the renewable energy message. There is some evidence that consumer recognition of the Green-e logo is increasing. The *Conscious Consumer Report* released by BBMG in 2009 indicates that 21% of consumers recognize the Green-e logo (Cook 2009). This is the fifth greatest recognition rate of conscious consumer logos, behind the recyclable logo, ENERGYSTAR® logo, and the USDA Organic logo.[6]

Broader research on whether consumers recognize or understand various statements or logos, however, is limited. Even the FTC in its review of the Green Guides commented that more research in this area is needed. In the consumer perception research that the FTC conducted, there was a fair amount of consumer confusion regarding what it means for a product to be "made with renewable energy." Respondents indicated, for example, that a "made with renewable energy" claim suggested that the product also was made with renewable materials (28%) or recycled materials (21%) (FTC 2010).

Text Box 5. Green-e Re:print Program

The proposed revisions to the FTC's Green Guides specify that marketers should clarify whether renewable energy claims apply to the product, the packaging, or a particular component of the product or packaging because it could be unclear to the consumer what the renewable energy claim refers to. For example, if a greeting card carries the statement "made with renewable energy," it is unclear whether renewable energy powered the offices where the content of the greeting card was developed, the facility that manufactured the paper, the facility that printed the card, or all three. In response to this confusion, Green-e has developed the re:print program specifically for paper and printing companies. This program allows companies working with a Green-e certified printer and using Green-e certified paper to use the Green-e logo on printed materials. The logo specifies that the paper is manufactured and printed using renewable energy (Center for Resource Solutions 2010).

For some companies, consumer understanding could be less of a concern. Rick Halverson of Great River Organic Milling, for example, commented that the company uses renewable energy because it believes in it, and that if consumers do not recognize or understand the Green-e logo on a product it is not a major concern (Halverson 2011). Companies with less concern centered on consumer understanding tend to be the ones using renewable energy labeling to confirm a commitment to sustainability.

Beyond consumer recognition and understanding is the basic challenge of whether the message will resonate with consumers and influence purchase decisions. Consumer survey research exists revealing that consumers prefer products made with renewable energy (Vestas and TNS Gallup 2011); it is unclear, however, whether this will translate into an actual increase in sales. Eco-products historically have suffered from inaction by consumers, even though surveys indicate a willingness to act (Rex and Baumann 2007). It is unclear whether products made with renewable energy see a similar gap between consumer preference and sales increases.

5.4. Product Packaging Costs

The costs associated with changing product packaging were cited as a relatively minor challenge. Companies tend to strategically time when to include the renewable energy messaging when a new product or product line is launched, when other essential changes to the product packaging were required, or when product-packaging contracts were being renewed. As a result, the additional cost of the renewable energy marketing message was minimal. Whenever a change is made to a product package, however, costs are involved. The process becomes increasingly more complex and costly as more products carry a message of being made with renewable energy.

Product packaging costs were identified as a greater challenge for companies in the business-to-business sector who have multiple layers of product packaging. Cascades Tissue Group, for example, noted that the costs associated with modifying both the inner and outer packages presented a considerable challenge. The medical technology company BD also brought up the challenge of products that contain multiple packaging layers, commenting that the different layers of packaging reach different audiences. For BD, labeling the outer package communicates with their business customer and labeling the inner package communicates with the end-user. If BD wants to communicate with both target audiences, product packaging labeling becomes a significantly greater challenge. Although BD currently does not useon-product messaging, it expects to begin labeling outer packaging in 2012 and to expand to inner packaging over time.

5.5. Cost of Certification and Program Requirements

When using a third-party certification program, the cost of certification and program requirements was identified as a minor challenge. As Donald Bozek of Strathmore Artists Paper put it, this is just a cost of doing business (Bozek 2011). One company did identify the certification costs to be a considerable challenge because of the way the certification program was structured. The program required a fee that was based on the company's total revenue even though the company only was using the logo on one product line. This company thought that that the certification fee was not aligned with the use of the logo; as a result, the certification costs were viewed as a barrier.

5.6. Products Marketed Internationally

For companies serving international markets there are additional challenges to renewable energy marketing. Perhaps the most basic challenge is satisfying varying definitions of what renewable energy is, as well as guidelines on the types of renewable energy marketing claims that can be made. A second challenge relates to consumer recognition, understanding, and relevance. A logo that is recognized by a consumer in the United States, for example, might be unrecognized in other markets.

A more practical challenge when selling a product in multiple markets is that the consumers likely speak different languages. Any renewable energy marketing messages using text might display text in many languages, further reducing already limited product real estate. Strathmore Artists Paper, for example, serves both Canadian and U.S. markets and, therefore, includes messages in English, Spanish, and French.

CONCLUSION

Marketers have highlighted a variety of environmental attributes of products for several decades, but specifying the type of energy used to manufacture a product is a trend that has only emerged in the past decade. In the United States, the number of companies marketing products as "made with renewable energy" has increased substantially in the past several years, although the market remains small. In 2011, researchers identified about 50 companies that use renewable energy labeling on products. Through interviews with 20 companies with experience labeling products as "made with renewable energy," we have found the following: (1) the companies range in size, (2) some companies are in the business-to-business sector and many others are in the business-to-consumer sector, (3) companies span a wide array of industries, and (4) companies vary in terms of the scope of their labeling efforts, with some labeling a specific product and others labeling a product line or all company product lines.

Motivations for labeling products differ. Some companies are motivated by the ability to differentiate a product through using a "made with renewable energy" message, appealing to the environmentally conscious consumer. Other companies reported that the primary motivations were to confirm an already existing commitment to sustainability or to enhance the image of a brand.

None of the companies interviewed attempted to earn a price premium on the product because of the use of renewable energy in its manufacture.

Methods of communicating the use of renewable energy vary. Some organizations have relied on logos or imagery to convey the use of renewable energy in the manufacture of a product and others have included text on the product packing. Others have complemented this by providing information on websites about their renewable energy purchases and the associated benefits, and some have used social media and other promotions to convey their use of renewable energy.

There are numerous challenges associated with communicating the use of renewable energy directly on products, including the limited real estate available on product packaging and determining the appropriate content to include. Another challenge is the difficulty in assessing how effective "made with renewable energy" claims are in increasing product sales. Further, there often is uncertainty as to whether the customer recognizes the logo used and understands the information being conveyed. Lessons from earlier green-marketing experience indicate that a label or logo that is recognizable, easy to understand, and trusted is more effective at conveying information to the consumer—as a result, they contribute to market growth (Rex and Baumann 2007). The limited experience with renewable energy logos, however, makes it difficult to adequately gauge recognition and understanding.

A multi-attribute label is something that could address multiple environmental attributes, not just use of renewable energy. Interviewed companies highlighted that a multi-attribute label could address concerns about limited packaging real estate and simplify the certification process on the company end.

Some lessons can be learned from how eco-products have been marketed; however, products that are made with renewable energy are unique in that they are identical to the comparable conventional product in quality and performance; only the energy used to manufacture the product is different. Given this new market, it remains to be seen the extent to which products "made with renewable energy" will be utilized and support the use of renewable energy. Corporate commitments to using renewable energy and communicating that commitment on product packaging and through other means could play an important role in educating customers about the availability and feasibility of using renewable energy as an alternative energy source.

APPENDIX A. SAMPLE OF LABELING EFFORTS

Company	Industry	Type of Labeling	Image
Aveda	Health & beauty, business-to-consumer	Green-e logo on website; description of efforts on website	
Boulder Canyon Chips	Food, business-to-consumer	Independent logo on product	
Buywell Coffee	Coffee, business-to- consumer	Green-e label on product	

Company	Industry	Type of Labeling	Image
Cascades Tissue Group	Paper, business-to-business	Green-e logo on product	
GNP Company	Food, business-to-consumer	Carbon Trust and Renewable Choice Energy logos on product	
Great River Organic Milling	Food, business-to-consumer	Green-e label on product	
HostPapa	Internet hosting, business-to-business	Green-e, Bonneville Environmental	

Appendix A. (Continued)

Company	Industry	Type of Labeling	Image
		Foundation, and Green Tags logos on website	
Intelligent Nutrients	Health & beauty, business-to-consumer	Green-e logo on product	
Kettle Brand	Food, business-to-consumer	Description of efforts on product	
Kraft (Philadelphia Cream Cheese)	Food, business-to-consumer	Independent logo on product	
New Belgium Brewery	Beer, business-to-consumer	Independent logo on product	 
New Leaf Paper	Paper, business-to- business	Independent and Green-e logos on website	

Company	Industry	Type of Labeling	Image
Santa Cruz Organics	Food, business-to-consumer	Green-e logo on product	
Sappi Paper	Paper, business-to- business	Description of Green-e certification on website	
SC Johnson	Household products, business-to-consumer	Statement and Independent logo	

Appendix A. (Continued)

Company	Industry	Type of Labeling	Image
Strathmore Artist Paper	Paper, business-to- consumer	Independent and Green-e logos on product	
WhiteWave Foods	Food, business-to-consumer	Bonneville Environmental Foundation label on product	

REFERENCES

Bennett, C. (4 August 2011). Phone communication. Aveda, Blaine, MN.

Berling, J. (12 August 2011). Phone communication. GNP Company/Just BARE Brand, St. Cloud, MN.

Bozek, D. (22 July 2011). Phone communication. Strathmore Artists Paper, Neenah, WI.

Bratter, D. (8 July 2011). Phone communication. WhiteWave Foods, Broomfield, CO.

Carbon Trust. (2010). "Pasta Becomes Latest Weapon in the Fight Against Climate Change." http://www.carbon-label.com/our-news/in-brief/pasta-becomes-latest-weapon-fight-against-climate-change. Accessed February 21, 2012.

Center for Resource Solutions. (2004). "Green-e Verification Report 2004." http://www.green-e.org/docs/VerificationReport04.pdf. Accessed January 9, 2012.

Center for Resource Solutions. (2010). "Green-e re:print Program Brochure." http://www.green-e.org/docs/CLP/reprint/rePRINTbrochure.pdf. Accessed January 9, 2012.

Center for Resource Solutions. (2011). "Green-e Verification Report 2010." http://www.green-e.org/docs/2010%20Green-e%20Verification%20Report.pdf. Accessed January 13, 2012.

Cook, O. (2009). "Leveraging the Green- e Logo: An Introduction to Green-e Marketplace." http://www.green-e.org/docs/CLP/presos/ 120909.pdf. Accessed January 12, 2012.

Federal Trade Commission. (2010). *Proposed Revisions to the Green Guides.* http://www.ftc.gov/os/fedreg/2010/october/101006greenguides frn.pdf. Accessed January 9, 2012.

Gallastegui, I. (2002). "The Use of Eco-Labels: A Review of the Literature." *European Environment* (*12*), pp. 316–331.

Halverson, R. (21 July 2011). Phone communication. Great River Organic Milling, Fountain City, WI.

Hanson, C. (2005). "*The Business Case for Using Renewable Energy*." World Resources Institute. http://pdf.wri.org/corporate_guide_7.pdf. Accessed January 16, 2012.

Lawson, D. (11 July 2011). Phone communication. SC Johnson, Racine, WI.

Natural Marketing Institute. (April 2011). *Consumer Attitudes About Renewable Energy*. NREL/SR-6A20-50988. Golden, CO: National

Renewable Energy Laboratory. http://apps3.eere.energy.gov/greenpower /pdfs/50988.pdf. Accessed January 9, 2012.

Ott, S. (26 July 2011). Phone communication. Cascades Tissue Group, Kingsey Falls, Québec, Canada.

Peattie, K. & Crane, A. (2005). "Green Marketing: Legend, Myth, Farce or Prophesy?" *Qualitative Market Research: An International Journal,* (*8:4*), 357–370.

Rex, E. & Baumann, H. (2007). "Beyond Ecolabels: What Green Marketing Can Learn from Conventional Marketing." *Journal of Cleaner Production,* (*15*), pp. 567–576.

Thompson, L. (7 July 2011). Phone communication. Sappi Fine Paper North America, Boston, MA.

Trudel, R. (2011). "*A Primer on Socially Conscious Consumerism." Network for Business Sustainability*. http://www.nbs.net/wp-content/ uploads/NBS-Consumerism-Primer1.pdf . Accessed January9, 2012.

Vestas, TNS Gallup. (2011). "*Global Consumer Wind Study 2011*." http://www.vestas.com/Admin/Public/DWSDownload.aspx?File=%2fFiles%2fFiler%2fEN%2fPress_releases%2fBloombergVestas_june_2011%2fGlobal_Consumer_Wind_Study_2011.pdf. Accessed January9, 2012.

Wallace, K. (21 July 2011). Phone Communication. New Belgium Brewing, Fort Collins, CO.

End Notes

[1] Members of the Green-e Marketplace include companies in industries such as food and beverage, health and beauty, printing and packaging, manufacturing, media and communications, Internet and hosting, home and office, art, life science, events, and non-profit. http://www.green-e.org/base/pl_products?dd_74=1&test=y. Accessed January 9, 2012.

[2] Renewable energy can be from onsite generation or through the purchase of Green-e Energy certified renewable energy certificates (RECs). The minimum renewable energy use for the company is based on a sliding scale depending on the company's annual electricity use. See http://www.green-e.org/getcert_bus_howto.shtml for more information.

[3] Other founding partners include The Global Wind Energy Council, WWF, UN Global Compact, The LEGO Group, PwC, and Bloomberg.

[4] Companies also promote renewable energy use to their employees, for example, through banners. BD has used the Green-e logo on parking lot banners and on cookies for a company party.

[5] Additionally, new products tend to experience a spike in sales regardless of the type of messaging used on the product.

[6] The Smart Choices logo also ranked higher than the Green-e logo; however, the Smart Choices Program was terminated in 2010.

In: Renewable Energy Products
Editor: Brett R. Harris
ISBN: 978-1-62417-565-7

Chapter 2

GREENING YOUR PRODUCTS: GOOD FOR THE ENVIRONMENT, GOOD FOR YOUR BOTTOM LINE*

Environmental Protection Agency

SECTION 1. INTRODUCTION

Many innovative businesses have successfully introduced "green" products in recent years. While many of these products have saved money, they have also reduced impacts on human health and the environment by using recycled or recyclable material (e.g., recycled paper products), reducing their energy usage (e.g., efficient washing machines that use less water and energy) or eliminating the use of toxic chemicals during their manufacture, for example. How do these business owners and product designers go about improving existing products or introducing new green products? What motivates them to do so? In providing insight into these questions, this guidance document aims to encourage you to identify opportunities to green your existing products or introduce new green products to your product line.

This guidance document focuses on opportunities to help your business reduce the impact on the environment, while saving you money. Two

* This is an edited, reformatted and augmented version of the Environmental Protection Agency, dated February 2002.

approaches for improvement are product improvement and process improvement (Figure 1).

The goal for these two approaches is to benefit your bottom line, while also benefiting the environment. They aim to improve your profitability in a variety of ways, as shown in the table below.

It is helpful to consider the product development process (Figure 2) when considering how to incorporate greening concepts. While greening concepts should be incorporated into all product development phases, special consideration should be given to the product's design – studies show that about 70% of a product's environmental impacts are determined by decisions made during the design stage.

Incorporate greening concepts into your product development process.

Table 1. Greening is Profitable

Increasing market share	Attracting green consumers by getting your products included in environmentally preferable purchasing programs.
Reducing operating costs	Increasing efficiency and reducing waste leads to better profit margins.
Less tangible benefits	Includes benefits such as increasing employee participation and morale, strengthening the company's image in the community.

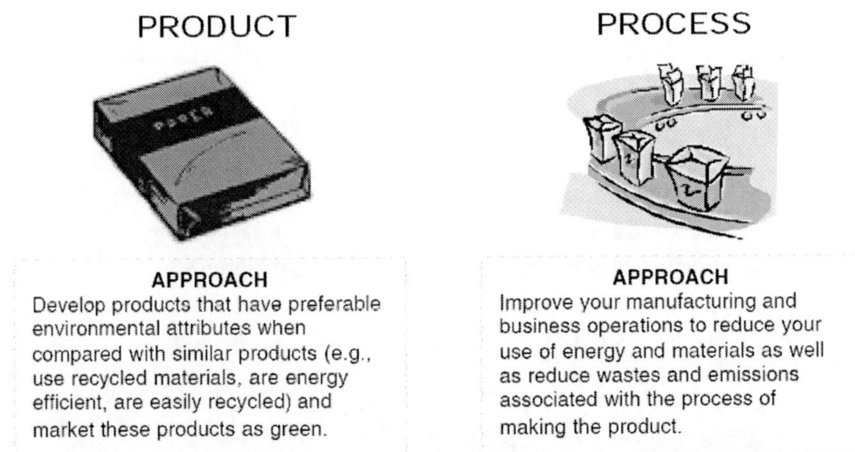

Figure 1. Product and Process Approaches to Greening.

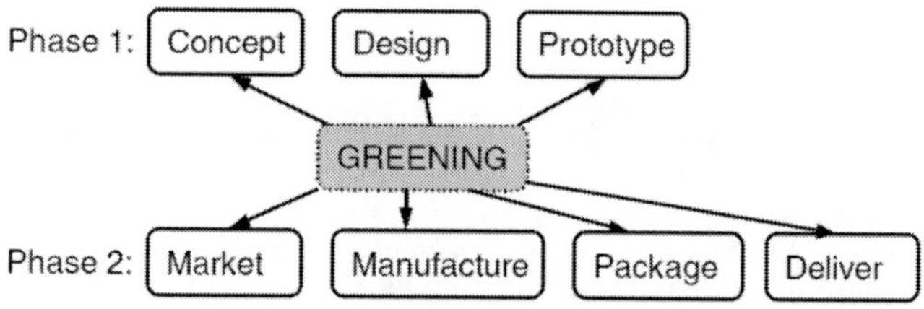

Figure 2. Incorporate Greening into All Development Phases of a Product.

Greening is a continuous process!

It is important to remember that the concept of "greening" is relatively new, and is therefore still a moving target. Government agencies and businesses alike are struggling to determine the best ways to improve their products to better protect human health and the environment. For example, greener materials not available today may be available tomorrow. Thus, greening is a continuous process, as new information and ideas become available to further improve your products and manufacturing processes. Therefore, identifying greening opportunities should be part of the standard product and process improvement practices of your organization.

Is There a Market for Green Products?

Federal agencies have been purchasing green products for quite some time, and are under increasing pressure to buy more of these products. As early as 1976, the Resource Conservation and Recovery Act directed federal agencies to buy recycled-content products. More recently, various Executive Orders as well as the Federal Acquisition Regulation (FAR) have expanded the environmental criteria required in making purchasing decisions (Table 2).

In addition, an estimated **$400 billion** is spent annually by state and local governments to procure products and services. A growing number of state and local governments are implementing environmentally preferable purchasing programs, including those shown below.

Each year the federal government procures more than **$200 billion** worth of products and services, giving these mandates the ability to significantly expand the market for green products.

Table 2. Recent Federal Green Procurement Mandates

Executive Order 13101 Greening the Government Through Waste Prevention, Recycling, and Federal Acquisition This Order requires federal agencies to consider environmental criteria in their purchasing decisions, states a preference for procuring "environmentally preferable" products and services, and directs the U.S. Environmental Protection Agency (EPA) to develop guidance for environmentally preferable purchasing (EPP). *For more information see:* http://www.epa.gov/oppt/epp/pdfs/13101.pdf
Federal Acquisition Regulations These are the rules governing how federal agencies buy goods and services. These regulations specifically require federal agencies to "implement cost-effective contracting preference programs favoring the acquisition of environmentally preferable and energy-efficient products and services" (FAR 23.704a). *For more information see:* http://www.arnet.gov/far/
Biomass Research and Development Act of 2000 This Act promotes the development and manufacture of biobased products. *For more information see:* http://www.bioproducts-bioenergy.gov/bio_act.html

Table 3. States, Counties, and Cities Implementing EPP Programs

States	Connecticut, Delaware, Indiana, Maine, Maryland, Massachusetts, Pennsylvania, Texas, Vermont, Washington DC, Wisconsin
Counties	California: Orange County, San Diego County Florida: Lee County Michigan: Kalamazoo County, Washtenaw County Minnesota: Olmstead County Missouri: Jackson County Nevada: Washoe County New Jersey: Cape May County North Carolina: Chatham County Oregon: Multnomah County Virginia: Fairfax County Washington: King County
Cities	Arizona: Phoenix California: Chula Vista, Oakland, Richmond, San Diego, San Francisco, Santa Barbara, Santa Monica Colorado: Boulder Illinois: Chicago Missouri: Kansas City New York: New York City Ohio: Cincinnati, Dayton Oregon: Portland Tennessee: Chattanooga Washington: Seattle

The above list will expand over time as more governments see the value in implementing greener procurement systems and learn from each other. As these procurement systems grow and become more integrated into standard practice, opportunities to successfully market your green products will be greatly enhanced.

Resource: Information About State and Local Governments

State and Local Government Pioneers: How State and Local Governments Are Implementing Environmentally Preferable Purchasing Practices. *November, 2000.*

http://www.epa.gov/opptintr/epp/pdfs/statenlocal.pdf

SECTION 2. WHAT IS A GREENER PRODUCT?

"Greener" products have a *lesser or reduced impact* on human health and the environment *than other products* that perform a similar function. Table 4 compares several traditional products to greener products.

Evaluating greener products generally requires a review of multiple environmental attributes. For example, in the above lighting example, while fluorescent bulbs use less energy than incandescent bulbs, these bulbs contain mercury that can be released into the environment if they are not properly disposed (e.g., not recycled).

Resource: Evaluating Multiple Environmental Attributes

This multimedia training tool provides an overview of the US EPA's Environmentally Preferable Purchasing program (EPP). Section IV: Environmental Performance Characteristics provides guidance on evaluating products across multiple environmental attributes.

http://www.epa.gov/oppt/epp/gentt/intver/app.html

Table 4. Examples of Greener Products

Product Category	Traditional Product	Greener Product	Improved Environmental Performance
Lighting	Incandescent bulb	Fluorescent bulb	Less energy used
Paper tablet	100% virgin paper	50% postconsumer paper	Reduces natural resource use
Copier paper	Chlorine bleached paper	Processed chlorine free paper	Reduces toxic emissions

Case Study: Recycled Writing Tablet
(Blind Industries and Services of Maryland)

100% recycled, 100% postconsumer, processed chlorine free writing tablet

Blind Industries and Services of Maryland (BISM) manufactures an extensive line of writing tablets. The Department of the Interior (DOI) approached BISM a couple of years ago about developing a writing tablet made from 100% recycled, 100% postconsumer, processed chlorine free paper. BISM agreed and collaborated with DOI to bring the product to market. DOI identified a vendor for the feedstock.

As the feedstock did not require any changes to BISM's manufacturing process, the new feedstock was easily incorporated into the new product. Introduced less than a year ago, the new tablets cost about 20% more than the standard 30% recycled content tablets.

Sales remain slow, but BISM is confident that just as with 30% recycled content paper, over time, product demand and production volume will increase, causing prices to drop and increasing the product's competitiveness.

Contact:
Paul Becker, Director of Marketing
Blind Industries and Services of Maryland http://www.bism.com
800-647-2468

Traditionally, product design takes into account standard product attributes such as price, performance, and safety.

To be competitive in the marketplace, green products must incorporate these traditional attributes. EPA's Environmentally Preferable Purchasing (EPP) program builds upon these traditional attributes by adding "Environment" and "Health" considerations into these traditional building blocks (Figure 3 on the next page).

This document focuses on two approaches for greening your products: greening the attributes of your products, and greening your manufacturing process (see Table 5). These two greening approaches are introduced below and then explained in greater detail in Sections 3 and 4 of this document.

Table 5. Two Approaches to Greening

Type of Greening	Examples
Greening product attributes	• Paper made from recycled content • Flatware that is biodegradable
Greening the manufacturing process	• Replacing traditional solvents with citrus-based solvents • Installing energy efficient motors

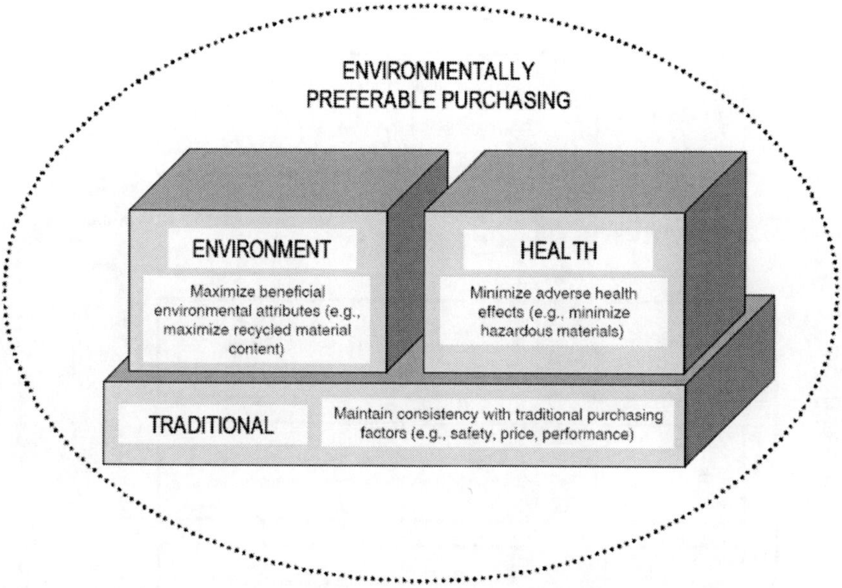

Figure 3. Characteristics of Environmentally Preferable Purchasing.

Overview: Greening Product Attributes

A product can be considered greener based on its characteristics, or "attributes." Those attributes can reduce the environmental impact of a product in a variety of ways. Examining a product's life cycle across its many stages provides a useful framework for thinking about how product changes will impact the environment. The diagram in Figure 4 below illustrates product attributes that can be improved for each life cycle stage, using paper manufacturing as an example.

Focus is usually placed on the environmental impacts arising from manufacturing. But it is also important to consider the upstream impacts (those that occur before the product is manufactured, such as mining or processing the input materials) and downstream impacts (after manufacture including distribution, use, and final disposal) of a product. When thinking of ways to improve a product's environmental effects, think about all stages of the product's life cycle. More detail on greening product attributes is provided later in this document.

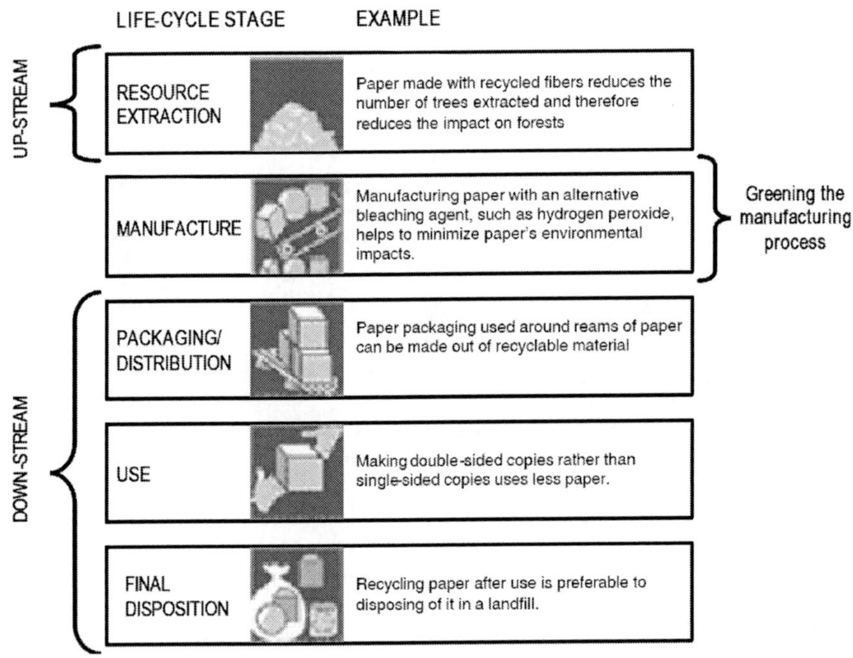

Figure 4. Life-cycle Diagram for Paper.

Case Study: Greening Plastic Flatware (Signature Works)

Signature Works, a division of LC Industries Inc., is a nonprofit company located in Hazelhurst, Mississippi that is affiliated with the National Industries for the Blind. Signature Works began work on a greener product to avoid the cost of large capital investments. These steps look likely to succeed in both avoiding costs and in ensuring the continuity and stability of their customer base.

Signature Works, a participant in the JWOD program, is the designated mandatory source to the government for plastic flatware (spoons, forks, knives etc.). Indeed, it is the principal supplier of plastic flatware to the armed forces and was commended by the U.S. Military for the quality of their supply support during the Gulf War. The division has approximately 500 employees and annual revenues of $25 million.

In February of 2000, the Defense Logistics Agency (DLA) announced that it intended to replace flatware made of polystyrene plastic with biodegradable and biobased alternatives and issued new descriptions and stock numbers for those items. To maintain its customer base and sole source position, Signature Works needed to radically modify its product and possibly its production process. For some businesses, making the changes might have generated a panicked rush. Signature Works, however, was not worried—the company was already working to develop a biobased biodegradable product.

Signature Works began assisting with the development of biobased and biodegradable plastic flatware in 1990, a decade earlier, when a supplier asked them to try using a resin made out of potato starch. Although that product proved unsuccessful for technical reasons, Signature Works realized that commercially viable biobased and biodegradable resins would ultimately be developed. If the company did not play an active role in the development of those resins, the resins that emerged as commercial products might well be incompatible with the company's manufacturing equipment. Replacing all of that equipment would be an expensive prospect—costing between $500,000 and $1,000,000. To avoid this eventuality the company sought out collaborations with Natick Army Research Center, polymer researchers, and resin suppliers, offering free use of its manufacturing facilities to test new and experimental resins.

Signature Works is currently testing various biobased and biodegradable resins. As of the end of 2001, resins tested thus far have been incompatible with Signature Works' equipment. However, the company will continue identifying and testing additional resins so as to maintain their supplier relationship to the US military.

Contact:
Vikki White
Signature Works
800-647-2468 vwhite@netdoor.com

Overview: Greening the Manufacturing Process

Products can also be made greener by improving the manufacturing processes involved in producing them. Several key characteristics of greener manufacturing are highlighted below. Some aspects of greener manufacturing translate into a product attribute (e.g., paper made with chlorine-free bleaching process). Many lead to cost savings through more efficient materials use, waste reduction and energy use reduction.

> Greening is good for employees.

Greening the manufacturing process also has positive impacts on human health of employees as well as the surrounding community. For example, traditional wall paints contain solvents, many of which are volatile organic compounds (VOCs.) VOCs emitted while manufacturing or using the product can contribute to the formation of ground-level ozone, a lung irritant that makes breathing difficult. Some paint solvents can also cause lung irritation in workers manufacturing or using paints. The paint manufacturing process can be made greener by reducing solvent use (i.e., by using solvents more efficiently in the manufacturing process) or by replacing the solvents with those that are less volatile and toxic.

Table 6. Characteristics of Greener Manufacturing

Consumes less natural resources
Requires less energy to produce
Releases less gases that lead to global climate change
Uses and releases less toxic materials
Creates less waste (both hazardous and non-hazardous)

Case Study: Printing Green – Small Changes, Big Dividends (Custom Print)

Custom Print is a sheetfed, offset lithographic printer of commercial color products including brochures, folders, and booklets. They are a 22 year-old company employing 30 people with $4 million in annual sales. For them, greening their process has been a profitable experience that took place by implementing small changes over the course of a few years. The following highlights several of their initiatives and resulting cost savings:

1) **Streamlining inventory:** As a result of process mapping (further described in Section 4), Custom Print examined its inventory and purchasing records, revealing that over 80 different chemicals were on-site. A team of press operators, purchasing staff and maintenance personnel investigating the causes for the large inventory suggested ways for reducing the chemical inventory. First, the team discovered that sometimes multiple chemicals were being purchased to perform the same task. Employees using these chemicals got together to agree upon one chemical that would be used for the task. Second, the team identified chemicals that could perform multiple tasks as a means to further reduce inventory. Lastly, the team recommended returning unused or partially used samples to vendors. After implementing these changes, Custom Print cut its inventory by 70% – to just 24 chemicals – and is saving an estimated $5,000 per year in inventory and waste disposal for expired chemicals, while reducing its potential liability.
2) **Modifying ventilation:** After an employee complained of headaches from isopropyl alcohol (IPA) in the fountain solution, Custom Print installed an air conditioning system with a high volume fan. A year later the company switched to an alcohol-free system, but the fan kept running until some time later when an employee accidentally turned it off.
   This accidental flip of the switch made the company realize that less air movement was needed in the shop with the new fountain solution. The change in ventilation proved to be very profitable – reduced air conditioning needs cut electricity costs by $2000/month (40%) and enabled the company to renegotiate its service contract, saving $200/month. Heat from the presses was now enough to keep the room warm in winter, cutting fuel costs by $400/year. The slower air circulation also improved the consistency of print quality by keeping temperature and humidity more constant.
3) **Reducing wasted ink:** Custom Print was wasting ink because less frequently used colors often expired before they could be consumed. Working with its ink vendor, the company reduced wasted ink by training employees to mix specialty colors from existing ink stocks.

Initiative	Annual Cost Savings
Streamlining inventory	$5,000
Modified ventilation	$26,800
Reduced wasted ink	$8,000
TOTAL:	Approximately $40,000 per year in savings

Custom Print's experience demonstrates that:

- Greening can improve your bottom line
- Greening often means improving process or operational changes unrelated to product manufacture
- Greening includes changes in general non-industry specific infrastructure and services, such as heating and cooling
- Employees can help make improvements if given the opportunity

Source: US EPA Design for Environment Program, *Lithography Project Case Study 2: Pollution Prevention at Custom Print*, EPA-744-F-96-001, Aug. 1996.

Contact:
Stu McMichael, President, Custom Print http://www.customprint.com
703-979-6670

SECTION 3. GREENING PRODUCT ATTRIBUTES

The most common approach to greening a product is improving one or more of its environmental attributes. Green products are oftentimes benchmarked in comparison to other products that perform the same function. For example, you can benchmark the recycled content of your paper products based upon (1) the average recycled paper content currently found in the marketplace; (2) guidelines (described below) specifying recommended recycled content; or (3) recycled content levels found in other paper products marketed as green. All three benchmarks can provide a recycled content target for your products. The figure below summarizes the steps toward greening a product by improving its attributes. Each step is subsequently detailed in this section.

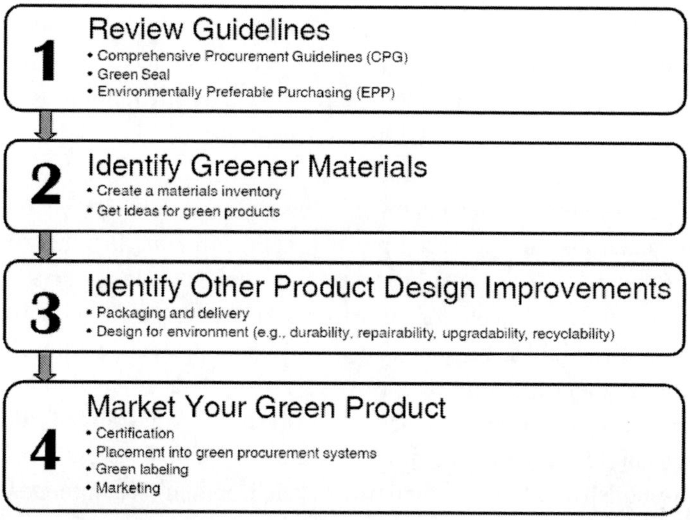

Figure 5. Product Approach to Greening.

Benchmark your green products.

Step 1: Review guidelines

A growing number of guidelines are becoming available to help purchasers select green products (Table 7). These guidelines can help you find the criteria your customers may use, help you develop criteria for greening your existing products, or provide ideas for new green products to introduce in your product line.

Comprehensive Procurement Guideline (CPG)

The federal government has undertaken various initiatives to include the environment in purchasing decisions. The Comprehensive Procurement Guideline (CPG) program is part of EPA's continuing effort to promote the use of materials recovered from solid waste. Buying recycled- content products ensures that the materials collected in recycling programs will be used again in the manufacture of new products.

The CPG program is authorized by Congress under Section 6002 of the Resource Conservation and Recovery Act (RCRA) and Executive Order 13101 ("Greening the Government Through Waste Prevention, Recycling, and Federal Acquisition," signed into law in 1998).

EPA is required to designate products that are or can be made with recovered materials, and to recommend practices for buying these products. Once a product is designated, procuring agencies (federal, state, or local agency using federal funds for procurement) are required to purchase the product with the highest recovered material content level practicable. A key component of the CPG program is EPA's list of designated products and recommended recycled content for each product. EPA has already designated or is proposing to designate the products listed in Table 8 (about 54 products). EPA also published final or proposed recycled content recommendations for each item. The CPG web site contains a brief description for each of the designated products, EPA's recommended recycled content ranges, and a list identifying manufacturers, vendors, and suppliers for each item. An example of a CPG guideline entry is shown in Figure 6.

Q: What is the difference between post-consumer and recovered materials?

A:

- Post-consumer material means a material or finished product that has served its intended use and has been diverted or recovered from waste destined for disposal, after having completed its life as a consumer item.
- Preconsumer material means waste material that is generated from, and commonly reused within, an original manufacturing process.
- Recovered material means both post-consumer and preconsumer materials.

Table 7. Guidelines at a Glance

Detailed on page 14	**Comprehensive Procurement Guideline (CPG)**
Sponsoring Organization:	U.S. Environmental Protection Agency
Key component:	Contains a list of designated products (and manufacturers)

Website:	and accompanying recycled content recommendations. http://www.epa.gov/epaoswer/non-hw/procure/
Detailed on page 17 **Sponsoring Organization:** **Key component:** **Website:**	**Green Seal** Nonprofit environmental labeling organization Provides a set of environmental criteria for products and certifies that a particular product meets these standards. http://www.greenseal.org/
Detailed on page 19 **Sponsoring Organization:** **Key component:** **Website:**	**Environmentally Preferable Purchasing (EPP)** U.S. Environmental Protection Agency The document "EPA's Final Guidance on Environmentally Preferable Purchasing" helps Executive agencies integrate environmental considerations into their buying decisions. http://www.epa.gov/oppt/epp/index.htm

Table 8. Current List of Products with CPG Requirements

Construction Products • Building insulation products • Carpet • Carpet cushion • Cement and concrete containing: a. Coal fly ash; b. Ground granulated blast furnace slag • Consolidated and reprocessed latex paint • Floor tiles • Flowable fill • Laminated paperboard • Patio blocks • Railroad grade crossing surfaces • Shower and restroom dividers/partitions • Structural fiberboard **Landscaping Products** • Garden and soaker hoses • Hydraulic mulch • Plastic lumber landscaping timbers and posts • Lawn and garden edging	**Paper and Paper Products** • Commercial/industrial sanitary tissue products • Miscellaneous papers • Newsprint • Paperboard and packaging products • Printing and writing papers • Park and Recreation Products • Park benches and picnic tables • Plastic fencing • Playground equipment • Playground surfaces • Running tracks **Transportation Products** • Channelizers • Delineators • Flexible delineators • Parking stops • Traffic barricades • Traffic cones **Vehicular Products**

Table 8. (Continued)

- Compost made from yard trimmings or food waste

Nonpaper Office Products
- Binders, clipboards, file folders, clip portfolios, and presentation folders
- Office recycling containers
- Office waste receptacles
- Plastic desktop accessories
- Plastic envelopes
- Plastic trash bags
- Printer ribbons

Toner cartridges

- Engine coolants
- Re-refined lubricating oils
- Retread tires

Miscellaneous Products
- Awards and plaques
- Industrial drums
- Mats
- Pallets
- Signage
- Sorbents

Manual-grade strapping

Note: Many mandatory source JWOD products already meet environmental purchasing criteria, including: clipboards, notebooks and binders; hanging file folders; paper towels and napkins; steno pads, pens and pencils.

EPA's Recommended Recovered Fiber Content Level

Item	Post-consumer Fiber (%)	Recovered Fiber (%)
Bathroom tissue	20-60	20-100
Paper towels	40-60	40-100
Paper napkins	30-60	30-100
Facial tissue	10-15	10-100
General purpose industrial wipers	40	40-100

NOTE: The content levels should be read as X% recovered fiber, including Y% post-consumer fiber and not as X% recovered fiber plus Y% post-consumer fiber.

List of Manufacturers and Suppliers

[There are 40 entries in this list; the following is a sample entry]

Javits-Wagner-O'Day Program
c/o National Industries for the Blind (NIB)/NISH
1901 North Beauregard Street; Alexandria, VA 22311
Phone: 800 433-2304; Fax: 703 998-5416

The Javits-Wagner-O'Day (JWOD) Program is a mandatory source for federal customers. JWOD nonprofit agencies furnish bathroom tissue with 100 percent recovered fiber and 20 percent post-consumer fiber content.

Figure 6 Sample CPG Guideline— Commercial/Industrial Sanitary Tissue Products.

Green Seal

Green Seal is a nonprofit standard setting organization that awards the "Green Seal of Approval" to products that cause less harm to the environment than other similar products as defined by their certification standards. Green Seal both develops environmental standards for product categories as well as carries out the actual certification of particular products from interested manufacturers. Table 9 lists currently available product certifications.

Q: What if I do not want to pursue Green Seal Certification?

A: The certification information provided by Green Seal is very valuable even if you choose not to pursue certification. It can give you ideas of green products that you may wish to develop, and to understand the generally accepted green attributes of that product.

For more information on certification, see the section below called "Marketing Your Product."

Table 9. Current List of Available Green Seal Standards

Tissue Paper	Dishwashers
Re-Refined Engine Oil	Cook tops/Ovens/Ranges
Compact Fluorescent Lamps	Watering Hoses
Water Efficient Fixtures	Alternative Fueled Vehicles
Printing and Writing Paper	Anti-Corrosive Paints
Household Cleaners	Plastics Labeling Systems
Paper Towels and Paper Napkins	Showerheads
Coated Printing Paper	Paper Products Used in the Preparation of Food
Paints	Residential Air Conditioning Systems
Windows	Fleet Vehicle Maintenance
Window Films	Powdered Laundry Bleach
Newsprint	Electric Chillers
Reusable Utility Bags	Residential Central Air-Source Heat Pumps
Refrigerators	Degreasers
Freezers	Food Service Packaging
Clothes Washers	Commercial Adhesives
Clothes Dryers	Industrial & Institutional Cleaners

Example of a Green Seal Standard: Tissue Paper (GS-01)

Second Edition, February 12, 1992 Copyright 1992 Green Seal, Inc.

A. Product Specific Performance Requirements
   1. Product must be made in accordance with reasonable industry practice with respect to holes, tears, wrinkles, cleanliness, foreign materials or dirt. It must have no disagreeable odor, either wet or dry, in accordance with reasonable industry practice. Edges of the product must be cleanly cut and not ragged. Product must dispense properly from dispenser or roll.
   2. Each roll of bathroom tissue must contain at least 40 square feet of product (300 x 4.5" x 4.4" sheets). Each box of facial tissue must contain at least 70 square feet of product (175 x 8.0"x 8.0" sheets).

B. Product Specific Environmental Requirements
   1. Bathroom Tissue: the fiber in bathroom tissue shall contain 100% recovered materials, including 20% post-consumer materials.
   2. Facial Tissue: the fiber in facial tissue shall contain 100% recovered materials, including 10% post-consumer materials.
   3. Post-consumer Content: the post-consumer content of a product shall be determined by measuring the average product fiber utilization over a period of no longer than three months.
   4. De-inking of Recovered Paper: recovered paper shall not be de-inked using a solvent containing chlorine, or any chemicals listed by the EPA under Section 313 of the Emergency Planning and Community Right To Know Act.
   5. Bleaching: If a bleaching agent containing chlorine is used, the adsorbable organic halogen content of the effluent from the production facility shall not exceed 1.0 kg per air dried metric ton of pulp. Chlorine and its derivatives shall not be used after January 1, 1996.
   6. Additional Ingredients: the product (not including packaging) shall not contain any added pigments, inks, dyes, or fragrances.

C. Packaging Requirements
   1. The core of a roll of bathroom tissue or the box used to package facial tissue must be made from 100% recovered fiber.
   2. A consumer package of bathroom tissue must contain at least 4 rolls unless the outer wrapper has been manufactured from 100% recycled materials.

3. Packaging shall not contain inks, dyes, pigments, stabilizers, or any other additives to which any lead, cadmium, or hexavalent chromium has been intentionally introduced.
4. The sum of the concentration levels of lead, cadmium, mercury, and hexavalent chromium present in any package or packaging components shall not exceed 100 parts per million by weight.

D. Labeling Requirements
1. The Green Seal Certification Mark may appear on the packaging and the product itself.
2. Whenever the Certification Mark appears on a package or product, the product or package must contain a description of the basis for certification. The description shall be in a location, style, and typeface that are easily readable by the consumer.

Reference: http://www.greenseal.org/standards.htm

Before a product gets the Green Seal, it must go through a certification process administered by Green Seal to ensure it meets the environmental standards that Green Seal has established. Unlike the Comprehensive Procurement Guidelines, which do not actually certify products from individual companies, Green Seal operates as an unbiased, third party certification organization.

After a product is successfully certified, it is permitted to display the Green Seal label. Having a Green Seal certification for your product increases its visibility and marketability.

Environmentally Preferable Purchasing (EPP)

The U.S. federal government is the single largest consumer of goods and services in the U.S., spending more than $200 billion annually on goods and services. EPA recognizes that this tremendous purchasing power influences what products and services are produced. EPA's Environmentally Preferable Purchasing (EPP) program, a federal program (established pursuant to Executive Order 13101) leverages that power by encouraging and assisting Executive agencies in purchasing environmentally preferable products and services.

EPP's goal is to empower federal consumers to make smart purchasing decisions that reflect environmental considerations. Various tools developed by the program assist this goal.

On-line EPP Tools
Use the "General Environmentally Preferable Purchasing Training Tool." This Interactive Training Tool is intended to help quickly and easily understand Environmentally Preferable Purchasing. http://www.epa.gov/opptintr/epp/finalguidancetoc.htm Additional information can be found in the document "EPA's Final Guidance on Environmentally Preferable Purchasing." http://www.epa.gov/opptintr/epp/finalguidancetoc.htm
Use the on-line "EPP Database" to learn more about green standards for specific products. http://yosemite1.epa.gov/oppt/eppstand2.nsf

Q: How is "environmentally preferable" defined?

A: Executive Order 13101 defines environmentally preferable as: "...products or services that have a lesser or reduced effect on human health and the environment when compared with competing products or services that serve the same purpose. This comparison may consider raw materials acquisition, production, manufacturing, packaging, distribution, reuse, operation, maintenance or disposal of the product or service."

General EPP Training Tool— This interactive training tool is intended to help you quickly and easily understand environmentally preferable purchasing. The tool includes:

- General overview, which covers basic EPP principles, including the definition of EPP and its environmental benefits.
- Current EPP mandates, including Executive Order 13101 and Federal Acquisition Regulations.
- EPA's Five Guiding EPP Principles that guide the process of incorporating environmental considerations into purchasing decisions.
- Environmental Performance Characteristics, which provides guidance on evaluating the environmental impacts associated with purchasing decisions across multiple environmental attributes.

- Procurement Policy, which describes how to incorporate environmental considerations into the federal purchasing process.
- Existing Green Purchasing Programs, describing various federal programs.
- Promising Practices Preview, describing successful EPP strategies that have been implemented by federal agencies.

The five guiding principles developed by the EPA provide broad guidance for applying EPP in the federal government setting. You should review these principles when considering the performance of your own products, and when providing information to consumers about your products. A summary of the principles is presented below.

Table 10. Five Guiding Principles for Environmentally Preferable Purchasing

Guiding Principle 1: Environment + Price + Performance = Environmentally Preferable Purchasing— Environmental considerations should become part of normal purchasing practice, consistent with such traditional factors as product safety, price, performance, and availability.
Guiding Principle 2: Pollution Prevention— Consideration of environmental preferability should begin early in the acquisition process and be rooted in the ethic of pollution prevention, which strives to eliminate or reduce, up-front, potential risks to human health and the environment.
Guiding Principle 3: Life Cycle Perspective/Multiple Attributes— A product or service's environmental preferability is a function of multiple attributes from a life cycle perspective.
Guiding Principle 4: Comparison of Environmental Impacts— Determining environmental preferability might involve comparing environmental impacts. In comparing environmental impacts, Federal agencies should consider: the reversibility and geographic scale of the environmental impacts, the degree of difference among competing products or services, and the overriding importance of protecting human health.
Guiding Principle 5: Environmental Performance Information— Comprehensive, accurate, and meaningful information about the environmental performance of products or services is necessary in order to determine environmental preferability.

EPP Database— Another useful resource is the Environmentally Preferable Purchasing (EPP) Database. This is a tool designed to make it easier for agencies to purchase products and services with reduced environmental impacts. The database is organized like a giant shopping mall with environmental information for selected products and services located within each store. Environmental information on over 600 products and services is included. This database links to the following information:

- Contract language and specifications created and used by federal and state governments and others to buy environmentally preferable products and services.
- Environmental standards and guidelines for the product you want to buy and vendor lists of product brands that meet these standards, where available.
- Other useful sources of information on the environmental preferability of products and services.

Step 2: Identify Greener Materials

Step 1 provides information and guidance for developing criteria for greener products. In this step you use this information to identify greener materials that can be substituted into your current products, or identify new green products to add to your product offerings.

Actions to Take
a. Create a materials inventory b. Get ideas for green products

a. Create a Materials Inventory— Whether you are considering greening an existing product or planning on introducing a new green product, a review of the input materials used in the manufacture of the product is necessary. The figure below shows a sample form that can be used to collect information on the various materials that go into a product. A copy-ready form is included at the end of this document.

The first column should contain the material and relevant specifications. The second and third columns should contain information on the current quantity and cost of the material. The fourth column should contain a list of the current vendors and alternate vendors that can be contacted. Use the appropriate trade association to get information on additional vendors that you may not have worked with in the past. EPPNET, described below, is also a useful resource.

The fifth column in the materials inventory sheet contains information about possible alternatives and goals based upon guidelines. For example, the alternative to ordinary plastic resin for flatware may be a biobased resin or a biodegradable resin. The alternative to ordinary paper napkins may be napkins with recycled materials. Green Seal standards require that napkins contain 100% recovered materials and at least 40% post-consumer material by weight— these specifications could be used as goals for your product.

Resource: Northeast Recycling Council (NERC) EPPNET

The Northeast Recycling Council (NERC) established EPPNET, the Environmentally Preferable Products Procurement Listserv, to link federal, state, local and private procurement and environmental officials charged with purchasing environmentally preferable products and developing policies for the procurement of these products.

The NERC EPPNET is intended to provide this group with quick access to information such as: availability of product specifications, lists of vendors for particular products, pricing information, strategies to achieve recycled product procurement goals, and federal procurement policies.

http://www.nerc.org/eppnet.html

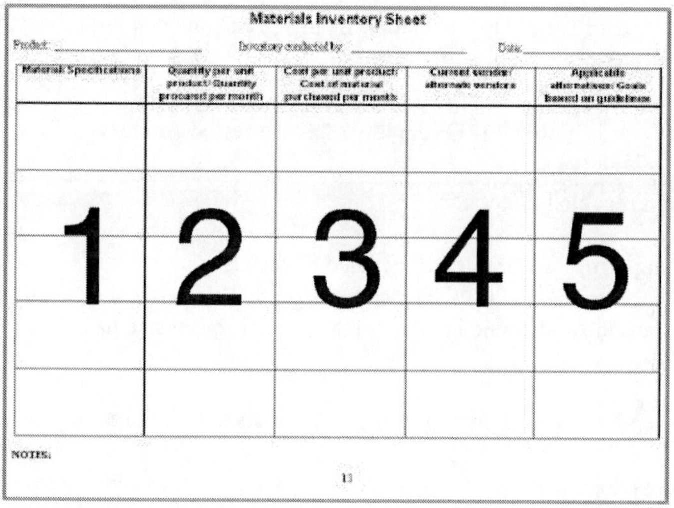

Materials Inventory Sheet

Product: ______ Inventory conducted by: ______ Date: ______

Material Specifications	Quantity per unit product/ Quantity procured per month	Cost per unit product/ Cost of material purchased per month	Current vendor/ alternate vendors	Applicable alternatives/ Goals based on guidelines
1	2	3	4	5

NOTES:

13

Figure 7. Materials Inventory Sheet.

b. Get Ideas for Green Products!— Use the following techniques to help you come up with ideas for greening existing products or introducing a new green product.

- **Green Guidelines**— Reviewing the guidelines presented in Step 1 (page 13) gives you an idea of what types of green products have penetrated the market to an extent that purchasing and certification programs have identified requirements and guidelines.
- **Market Survey**— Some products may not be mentioned in any of the guidelines, but are available as a green product. It is important to conduct a survey of the market to see if competitors are offering green products that serve the same function as your product. Characteristics of a competitor's product can be used as a guide for the greening of your own product. The two highlighted resources below (page 25) can be used to see what green products are available for purchasers and who currently manufactures or markets them.
- **Green Materials**— The table below describes various types of materials that can be used to make your product greener. Use this list in conjunction with the materials inventory sheet to identify opportunities to replace existing input materials with greener materials.
- **Contact vendors**— Contact your existing material vendors as well as new vendors to discuss the availability of greener materials. EPPNET is a useful resource for identifying green materials and vendors (see page 23).

Table 11. Overview of Greener Materials

Made from recycled materials: • Replacing virgin material with recycled materials. • Post-consumer recycled (PCR) content is preferable because it helps keep waste materials out of the landfill (preconsumer mill waste has always been recycled). Greener paper stock and inks for printed materials and packaging: †   • Tree free paper stock (e.g., sugar cane waste, straw, seaweed, algae and hemp). Alternative renewable paper sources can reduce need for wood pulp from old growth forests.

- Recycled paper stock that has not been de-inked. De-inking is an energy intensive process which may still result in toxic waste ink.
- Unbleached or nonchlorine bleached paper stock. Paper bleached with chlorine produces toxins such as dioxin, which are harmful to humans, fish and wildlife. Processed chlorine free (PCF) recycled paper is made using no chlorine during manufacturing. (However, some of the waste paper being recycled may have been bleached in the past with chlorine, so it can't be labeled totally chlorine-free.)
- Vegetable based printing inks such as soy inks. Vegetable based inks are renewable and emit less toxic volatile organic compounds (VOCs).
- Avoid the use of ink colors that contain high levels of heavy metals such as lead, copper, chrome, etc.

†Society for Responsible Design http://www.green.net.au/srd/

Less toxic or non-toxic:

- Non-toxic materials do not have the capacity to produce personal injury or illness to humans or other species through ingestion, inhalation, or skin absorption; less-toxic materials reduce such impact.
- Aim to replace a toxic material with a less toxic or non-toxic substances.
- Refer to Material Safety Data Sheets (MSDSs) to assist in determination of toxicity.

Low volatile organic compounds (VOCs):

- VOCs evaporate easily (volatilize) at room temperature and often have a sharp smell. They can come from many products, such as office equipment, adhesives, carpeting, upholstery, manufactured wood products, paints, solvents, pesticides, and cleaning products. Some VOCs can cause cancer in certain situations, especially when they are concentrated indoors. In the presence of sunlight, VOCs create ozone, a harmful air pollutant to both people and plants.
- It is preferable to use low-VOC versions of products to reduce risks to human health and the environment.

Biobased:

- Current biobased products and bioenergy technologies have the potential to make renewable farm and forestry resources major sources of affordable electricity, fuel, chemicals, pharmaceuticals, and other materials.
- They also have the potential to reduce dependence on foreign oil, improve air quality, water quality, and flood control, decrease erosion, and help minimize net production of greenhouse gases.
- Example is using soy-based inks for printing.

Table 11. (Continued)

Biodegradable:

- Recent ASTM Standards require that any product claiming to be biodegradable must completely decompose into carbon dioxide or water, the byproducts of decomposition, within a 180-day period.
- Note that the earlier generation of products described as biodegradable would often fragment and breakdown into big pieces that would not completely biodegrade.

Other considerations:

- Materials that will reduce hazards posed to workers when handled during product manufacture.
- Materials that can be obtained locally will reduce the environmental impacts of transportation.

- Materials that are easier to work with or require simpler processing.

Resource: "Environmentally Preferable Purchasing Guide"

Minnesota Solid Waste Management Coordinating Board

This Guide is a reference tool for government and school purchasers in MN who want to make wiser purchasing choices. This Guide helps them find ways to reduce workplace hazards, consume less energy, and protect natural resources by identifying environmentally preferable alternatives in over 30 product areas. It also provides information on cost and performance and sources of these products.

http://www.swmcb.org/EPPG/1_1.htm

Resource: "GreenOrder.com" Website

GreenOrder is a new resource for government institutional buyers and suppliers of energy efficient, recycled, and other environmentally preferable products. GreenOrder enables buyers to source and submit "Request for Quotes" (RFQs) for a broad range of products provided by suppliers registered with GreenOrder.

http://www.greenorder.com/home

Case Study: Recycled Trash Bags (Envision, Inc.)

100% recycled-content (30% postconsumer) trash bags

Envision, Inc. in Wichita, Kansas is the largest supplier of trash bags through NIB. To leverage increasingly "green" purchasing preferences of the federal government, Envision decided to introduce a recycled-content plastic trash bag. Although the typical recycled content in plastic products is 30% or less, Envision decided to develop a 100% recycled, 30% postconsumer content product, believing this would give them a strategic advantage over competitors. Product marketing reflects this choice and environmental attributes are emphasized as the defining feature of the product. To locate a feedstock, Envision simply requested the desired recycled content from their suppliers. Production of the new bags required no significant process alterations. The recycled-content bags must be made from somewhat thicker plastic film in order to perform as well as the virgin plastic product. But, because the recycled-content film also costs approximately 15% less than virgin stock, the final product is price competitive. Added to the procurement list in 2000, the sales of bags have increased each year since the product was introduced and rose 10% in the last year alone. Government clients currently account for 90% of sales.

Contact:
Envision, Inc. http://www.envisionus.com/
316-267-2244

Case Study: Biodegradable, Biobased Plastic Bags (Multi-Community Diversified Services Inc.)

Plastic trash bags are the bane of large-scale composting operations that handle residential yard waste today. Communities and contractors usually feel compelled to permit residents to use polyethylene or PVC bags because of their convenience and resistance to moisture.

But accepting such bags is costly – contents must be debagged, the final product is invariably contaminated with plastic fragments lowering its quality, and an average of 20% of the input is wasted in contaminated "overs" (i.e., contaminated product that must be disposed of). All this adds up to bottlenecks and slower throughput, lower prices, lost sales revenue, and significant disposal costs.

For a facility processing 15,000 tons, which might serve anywhere from 15-45,000 households depending on the region, the total cost could run anywhere from $184,000 – $364,000 depending on disposal fees and compost market prices. Multi-Community Diversified Services Inc. (MCDS), a NISH affiliate based in McPherson, Kansas, has a product that can substantially reduce these costs—a biodegradable plastic bag.

Founded in 1974, MDCS provides employment and support services for people with disabilities in McPherson County (population 28,000). An annual budget of approximately $9 million supports a staff of 210 engaged in support services and manufacturing activities. MDCS offers both employment and residential support services. The organization provides job placement and on-the -job support. The organization manufactures wooden pallets, remanufactures toner cartridges (this operation alone accounts for over 25% of annual revenues), and does subcontract work for local businesses, such as cutting and packaging lengths of drip irrigation tubing. MCDS also runs a small yard waste drop-off and on-site composting facility.

While attending a recycling and composting conference, MCDS staff were introduced to Mater-Bi, a biobased biodegradable plastic resin made from cornstarch and polylactic acid manufactured by Novomont. Unlike many competing products, the biodegradability of plastic bags and other products made from Mater-Bi had met ASTM and international standards for biodegradation under a variety of conditions in third party testing. Noting that Mater-Bi bags sold in the US were manufactured in Germany and considering its own unique access to the federal market through the JWOD program, MCDS decided to explore the viability of setting up a US operation for manufacturing Mater-Bi biodegradable plastic bags. After ensuring it would have access to the resin, MCDS performed due diligence research on the technology and company and hired a professor from Wichita State University to conduct a formal financial analysis.

(As a rule MCDS seeks outside advice and assessment for large projects such as this one so that the Board of Directors can be confident that the proposal is sound).

MCDS will soon be moving into a new facility in which 3800 square feet have been set aside for bag production. The organization intends to phase in production elements over time. MCDS will begin by simply repackaging bags manufactured in Germany, then, as sales volume increases, move to manufacturing bags from imported Mater-Bi plastic film and finally to blowing its own plastic film from imported resin.

Initially, bags will be produced primarily for federal customers, such as military bases composting food waste and grounds keeping operations composting yard trimmings. MCDS is projecting revenues of $1.25 million on sales of 4.25 million bags in year one. MCDS is already repackaging bags for sale to government and commercial clients while waiting for final approval from the Committee for the Purchase from the Blind and Severely Disabled. MCDS is convinced that the potential market for biodegradable and biobased bags is enormous—half of all US States already restrict disposal of yard-waste in landfill—and it is determined to get in on the ground floor.

Contact: Barry Adamson, Executive Director, Multi-Community Diversified Services Inc. (620) 241-6699 mcds01@midusa.net

Step 3: Identify Other Product Design Improvements

There are other ways to green your products that will improve their environmental performance when used by the consumer (e.g., reduced product packaging) and finally disposed of (e.g., designed for recycling).

Areas of improvement include:

- Packaging/ delivery
- Durability
- Repairability/upgradeability
- Design for recycling/disassembly

Energy Star®

Energy Star was introduced by the US Environmental Protection Agency in 1992 as a voluntary labeling program designed to identify and promote energy-efficient products in order to reduce carbon dioxide emissions. EPA partnered with the US Department of Energy in 1996 to promote the Energy Star label, with each agency taking responsibility for particular product categories. Energy Star has expanded to cover most of the buildings sector, residential heating and cooling equipment, major appliances, office equipment, lighting, consumer electronics, and other product areas.

This approach helps improve your company's position for future trends in environmental improvement, oftentimes at no additional cost.

- Packaging/delivery— Greener packaging can have a significant impact on reducing the solid waste associated with the product when it reaches the consumer.

Greener Packaging Goals

- Eliminate materials
- Reduce or replace most harmful materials
- Minimize number of levels of packaging
- Minimize mass of each level
- Returnable/ reusable/ refillable/ consumable
- Recyclable

It is also important to consider the impacts associated with the delivery of products. The mass and volume of the shipped product will impact transportation costs and energy use. For example, concentrated cleaners can deliver the same results using a smaller quantity. The smaller quantity means smaller containers that can be made with less material, saving you money and reducing the amount of waste that your customer must dispose of (thereby saving them money too). The smaller containers also decrease transportation costs.

- Durability—Products that are more durable last longer. Therefore, fewer products are required, which reduces impacts on the environment. For example, toner cartridges that are intended to be refilled many times should be built to be more durable. Even if its initial cost to the consumer is higher, it will pay for itself since the consumer purchases cartridges less frequently.
- Repairability and Upgradeability— Products that can be easily repaired or upgraded (e.g., easily replaceable parts) are usually better from an environmental viewpoint. Computers require upgrading every few years. Instead of having to replace the entire computer, they can be designed such that the outer case and other elements can be retained, and upgraded parts (such as a new processor or CD-ROM drive) can be replaced.

- Design for Recycling/Disassembly— When appropriate, products should be designed to be easily recycled after use. For example, portable phones should be designed to allow the user to easily access and remove the rechargeable battery for recycling. The easier it is for the consume r to recycle, the more likely that recycling will occur. In addition, products designed to be easily disassembled facilitate the recycling process by allowing different materials to be more easily separated.
- Product Take-back (future)— Think into the future! It is expected that manufacturers of larger products (e.g., white goods, cars, appliances, tools) will be expected to take these products back from the consumer. Products should be designed so they can be easily disassembled, remanufactured or recycled.

Design for Recycling Goals

- Use fewer types of materials
- Choose materials for which there are currently available recycling programs
- Properly label the product as to which parts are recyclable
- Design for disassembly

Step 4: Market your Green Product

While many of your current marketing techniques are also applicable to green products, the environmental attributes of a green product differentiate your product in the marketplace. Actions you can take to market your green product are described below.

Actions to Take

a. Get your product certified
b. Get your product included in green procurement systems and lists
c. Appropriately label your green product
d. Use marketing strategies

a. Certification— There are currently several green certification programs in the U.S. Many are for specific product areas (e.g., forestry products), but others, such as Green Seal or Scientific Certification Organization, have a broader list of certifiable product categories. In the future, additional certification programs may become available.

Examples of Certification Programs

Green Seal: Certifies a wide variety of products.
http://www.greenseal.org

Scientific Certification Organization: Certifies a wide variety of products.
http://www.scs1.com

Smartwood: Certifies wood products.
http://www.nwf.org/northeastern/resources/guidelines.html

Green Seal

If you have a product for which a certification program is already established by Green Seal, you can follow the procedure outlined below to obtain certification. If you manufacture a green product that is not included, consider requesting Green Seal to add it to their program. (See page 17 for a discussion of Green Seal standards.)

An advantage of having Green Seal certification is the listing of your product in their "Office Green Buying Guide" and in their periodic publication "Choose Green Reports". These services are part of Green Seal's Environmental Partners Program that provides green buying assistance to organizations (businesses, government, academia) concerned about the environmental impacts of their purchases. Green Seal does the work for them by identifying the greenest products available. Green Seal's product recommendations take into account environmental impacts, product performance and packaging. There are currently over 450 member companies providing products certified by Green Seal. This offers the potential for wide visibility of your product directly with purchasing agents.

Table 12. Green Seal Certification Procedure

1. Review the Green Seal standard for your product category. (If a Green Seal Standard does not exist for your product, call Green Seal to discuss developing one.) Call Green Seal if you do not have a copy of the standard or if you or your technical staff has any questions. (For a listing of standards, see page 17.)
2. Identify the product or products you wish to submit to Green Seal for certification testing.
3. Call Green Seal to discuss the costs associated with evaluating your product for Green Seal certification. These costs reflect the nature and number of lab tests required and the cost of a site visit. At the end of this discussion, Green Seal will provide a firm quote of the costs involved in evaluating your product and conducting annual monitoring to ensure its continued compliance with the standard.
4. Sign Green Seal's Application for Certification and pay the product evaluation fee specified in Step 3.
5. Submit product samples for any required tests and schedule any required site visits. If the product meets the standards set forth in the standard, Green Seal will grant you the right to use the Green Seal Mark on the product, its packaging, and in product advertising.
6. Should a product fail to meet the standard, Green Seal will inform you of the reasons and, if you wish, work with you to make the changes needed to meet the standard. Green Seal keeps confidential all test results and the fact that you have submitted products for testing.
7. Work with Green Seal staff on the design of the certified product's packaging to include the Green Seal Mark and the accompanying text that spells out the environmental preferability of your product.
8. Green Seal begins annual monitoring to ensure the product continues to meet the Green Seal standards.

Scientific Certification Systems (SCS)

Scientific Certification Systems (SCS) was established in 1984 as the nation's first third-party certifier for testing pesticide residues in fresh produce. In the past 15 years, the company has evolved to become a testing and certification organization, evaluating a wide variety of food safety and environmental claims.

The company's environmental division certifies a wide variety of claims related to environmental performance in the product manufacturing and natural resource extraction sectors. Specific product attributes, such as recycled content and biodegradability, are certified under the Environmental

Claims Certification program (see Table 13 below for additional examples). SCS also certifies broader claims of environmental preferability based upon the product's life cycle. These certifications are specifically intended to establish important facts about the products that may be of particular interest to purchasers. Certifications are consistent with international standards. SCS consults with the US Federal Trade Commission (FTC) to clarify questions regarding claims and ensures that its claims comply with the FTC and relevant state environmental marketing guidelines.

b. Procurement Systems— The CPG program maintains a list of manufacturers, suppliers and vendors of EPA designated items. If your organization produces a designated item containing recovered material within the content ranges recommended by EPA, you can request to be added to the lists of suppliers. You will be asked to include information that substantiates the recovered materials content of your product.

Table 13. Examples of Attributes That Have been Certified By Scientific Certification Systems (SCS)

• Recycled content	• Recovered content
• Salvaged wood from urban sources	• Biodegradability
• No ozone depleting chemicals	• No VOCs / low VOCs
• No added formaldehyde emissions	• Organic ingredients
• Poison-free/ alternative to poison	• Water efficient

Table 14. Scientific Certification Systems (SCS): Certification Procedure

1. **Work Authorization:** SCS reviews initial information to determine the feasibility of certification. Applicant then enters into a Certification Agreement.
2. **Data Review:** Applicant discloses data to SCS. All disclosures are confidential. SCS evaluates these data, and requests additional information as needed. Data review includes upstream suppliers where appropriate.
3. **Claim Verification:** SCS certification engineer conducts an audit, typically including on-site inspection of manufacturing facilities and records, including upstream suppliers where appropriate. Product testing is also conducted where necessary.
4. **Certification:** If product claims are substantiated, SCS issues a certificate and authorized certification artwork as appropriate to the certification recipient.
5. **Monitoring:** Certified company submits updated data on an annual basis to SCS to ensure that the certified claim remains valid.

EPA is also interested in learning about products for potential future designation. Should you consider requesting that a product be added to the CPG program, consult the September 20, 1995 *Federal Register* notice describing EPA's information needs and decision-making process (http://www.epa.gov/epaoswer/non-hw/procure/pdf/1995rg.pdf).

c. Green labeling— It is important to label your green product in a way that it will attract consumer attention, but it is critical not to make unsubstantiated or misleading claims. The Federal Trade Commission (FTC) has set forth guidelines for the labeling and marketing of green claims on products.

FTC Guidelines Principles

1. Clear and prominent qualifications and disclosures.
2. Distinguish between attributes of product, package and service.
3. Do not overstate attribute.
4. Basis for comparisons must be clear.

FTC's guidance document, "Complying with the Environmental Marketing Guides," can help you properly label and market your product.

Table 15. EPA's CPG Program

EPA evaluates six primary concerns when examining products for designation:
1. Use of Materials Found in Solid Waste— All products must be manufactured with materials that are recovered or diverted from the solid waste stream. Particular attention is paid to those items produced from materials that are a significant component of the waste stream.
2. Economic and Technological Feasibility and Performance— EPA conducts research to verify that the product containing recovered materials is available and that the product meets industry or government specifications and performance requirements. It also is important for the product to be priced competitively and for there to be adequate competition among suppliers of the product.
3. Impact of Government Procurement— EPA considers the potential impact that government purchasers might have in minimizing the solid waste stream if they purchase a product made with recovered materials. For example, although widgets might be made from 100% post-consumer materials, if the government does not buy widgets, EPA is not likely to designate them because designation would not contribute to a reduction in the solid waste stream.

Table 15. (Continued)

EPA evaluates six primary concerns when examining products for designation:
4. Availability and Competition— The items EPA selects for designation are available from national, regional, or local sources. The relative availability of an item influences the ability of a procuring agency to secure an adequate level of competition when procuring it.
5. Other Uses for Recovered Materials— EPA also considers the possibility of one recovered material displacing another recovered material as feedstock, thereby resulting in no net reduction in materials requiring disposal; the diversion of recovered materials from one product to another, possibly creating feedstock shortages for one or both products; and the ability of manufacturers to obtain recovered materials in sufficient quantity to produce the item under consideration.
6. Other Considerations— The price of recycled-content items are generally cost-competitive with virgin products. Prices may be influenced, in part, by such factors as the quantity of a product purchased, transportation costs, etc. These factors exist whether the item is virgin or recycled. In addition, EPA does not intend to designate experimental or developmental products until it can be shown that they meet all of EPA's selection criteria.

The following is an example of a misleading claim:

- Paper is advertised as "environmentally friendly": Although it contains recycled fiber, the manufacturing process still produces and releases toxic materials. Therefore, the environmentally friendly claim is misleading.

The following is an example of an acceptable claim:

- Paper claims to be "30% recycled fiber": Some of the recycled fiber is post-consumer, the rest is paper manufacturing scrap never sold to customers. This claim is acceptable because it does not overstate the attribute.

Resource: FTC Guidelines On-line

"Complying with the Environmental Marketing Guides"
http://www.ftc.gov/bcp/conline/pubs/buspubs/greenguides.pdf

"Guides for the Use of Environmental Marketing Claims (Green Guides)" www.ftc.gov/bcp/grnrule/guides980427.htm

d. Use Marketing Strategies – Marketing your green product adds a whole new dimension to your traditional marketing approaches. Thus, it is important to consider these basic concepts:

1. **Understanding the green consumer mindset** – Consumer concern about the environment has steadily increased over the past two decades.
2. **Identify opportunities** – The demand for greener products exists and will grow in the future.
3. **Highlight direct benefits** – Consumers primarily buy products for direct needs, not to “save the planet”, and consumers purchase out of self-interest (e.g., top environmental concern is health).
4. **Don’t marginalize products by “green washing”** – Highlight functionality first and environmental benefits second.
5. **Educate and empower consumers** – Consumers need information to become more sophisticated about buying green.
6. **Maintain credibility** – Do not make unsubstantiated claims.

Resource: Strategic Green Marketing

The six strategies presented above are from: “Strategic Marketing of Greener Products,” by Jacquelyn Ottman and Virginia Terry http://www.greenmarketing.com/articles/JSP1Apr98.html
For additional articles: http://www.greenmarketing.com/Articles_and_Speeches.html

Case Study: Green Copy Paper
(Louisiana Association for the Blind)

Product Line Extension-100% recycled, 50% postconsumer processed chlorine-free copy paper

On January 19, 2001, EPA Headquarters (HQ) implemented a new policy requiring that copy paper used at HQ contain at least 50% postconsumer content rather than the Federal standard of 30% postconsumer content. The policy also specifies that processed chlorine free (PCF) paper should be used whenever available. The new requirements took effect at the Headquarters level first, with all Regional offices and labs to make the conversion by the end of October 2001.

In response, the Louisiana Association for the Blind (LAB) in Shreveport immediately began working to extend their product line to include a new copy paper product that would meet the EPA specifications.

It was an intensive, demanding process for both LAB staff and its suppliers. LAB had produced 100% recycled, 30% postconsumer paper since 1995. Due to the excellent relationships between LAB, its suppliers and the paper mills, development of the new product was a collaborative effort.

The first and most difficult step was finding a reliable source to produce the paper and ensure that it would meet all performance requirements of the finished product.

The new paper underwent extensive testing and had to be certified as 100% processed chlorine free by the producing mill. During the conversion process from roll stock to cut-size sheets, LAB performed further tests for performance in high speed copiers, laser printers and ink jet printers.

In July of 2001, LAB began actively marketing to the EPA Regional Offices and quickly reached a blanket purchase order agreement with Region 4. The new copy paper costs more than the 30% recycled-content paper, but is priced competitively for a specialty product.

Paper prices can fluctuate dramatically, but as demand increases, LAB expects the new 100% recycled, 50% postconsumer content copy paper product to remain within a price range attractive to the customer.

While the EPA is the only agency currently requiring the new paper, LAB anticipates that other customers will similarly change their specifications and LAB intends to be well positioned when demand increases.

As demand for the product grows, LAB expects to introduce other paper products with the same environmental specifications. Always thinking ahead, the organization has already completed additional research on 100% recycled, 100% postconsumer PCF paper products and currently has copy paper available to also meet that specification.

Contact:

Shelly Taylor, VP Ma rketing, Louisiana Association for the Blind, http://www.labbsciee.com/

318-635-6471 jabowen@softdisk.com

SECTION 4. GREENING THE MANUFACTURING PROCESS

Products can also be made greener by improving the processes used in their manufacture. Toxic materials used in manufacturing can be replaced with less toxic or non-toxic alternatives (e.g., substituting traditional solvents with citrus-based solvents). Materials can be recycled on-site so you reduce your material use (e.g., on-site recycling of used solvents reduces the need for solvent disposal while reducing your solvent purchases).

Upgrading your equipment may improve process efficiencies, enabling you to reduce your use of energy (such as installing energy efficient motors) and raw materials. These approaches are also commonly called pollution prevention (P2) or source reduction. Studies have shown that companies often overlook some of these money saving process improvements.

Two principles guide the approach to greening manufacturing processes:

Principle #1

Maximize the Percent of Input Material That Leaves the Facility as a Saleable Product

- Process improvements
- Improved housekeeping
- Better inventory control
- Segregate waste streams
- Identify ways to turn wastes into saleable products to other companies
- Recycling process wastes

Principle #2

Minimize Use of Energy and Ancillary Materials

- Energy and water conservation
- Recycling programs
- Process optimization
- Regular equipment maintenance
- Employee training and incentive programs

This section helps you identify opportunities to green your manufacturing processes by using materials and energy more efficiently, while also saving money.

Improving the use of resources during manufacturing results in cost savings through:

- Reduced input materials
- Reduced waste
- Reduced emissions
- Reduced energy consumption
- Improved worker safety

The following diagram summarizes the steps you can take to identify opportunities for greening your manufacturing processes. Each step is subsequently detailed in this section.

Figure 8. Steps in Greening the Manufacturing Process.

Step 1: Map Your Processes

Understanding your manufacturing processes is the first step in identifying greener manufacturing opportunities. It is usually assumed that we understand how the industrial processes work in our facility, but oftentimes we do not really understand exactly how processes work or whether they can be improved, simplified or eliminated. Many times the person responsible for disposal of wastes does not understand the systems that are generating those wastes, or personnel continue to conduct tasks as they always have, without really understanding certain aspects of the function of these tasks. This can lead to the inefficient use of resources, such as using more solvent than is necessary or running water through a rinse system at too high a flow rate. Process mapping helps start the process of improving operations at your facility by creating an outline that gives you and your team a better understanding of how things work. What is a process map? A process map is a schematic description of a process or processes. It is not a quality improvement flowchart, a floor plan, nor a complex wiring diagram. Enough detail should be included to provide adequate information to understand the bigger picture of your processes and stimulate discussion on the interactions between various processes.

Reference for this section: Pojasek, Robert; "Understanding a Process with Process Mapping," Pollution Prevention Review, Summer 1997; p. 91-101.

http://www.pollutionprevention.com/Pdfp2/ understanding-a-process-with-process-mapping.pdf

Process maps are usually prepared by a team rather than by one individual because individuals with specific skills and knowledge of subsystems in the manufacturing process need to come together and make these connections. A completed process map should be the result of the integration of these subsystems into a map that allows you to look at the whole picture.

Benefits of Process Mapping

As a result of process mapping, Custom Print, a lithographic printer, realized that over 80 different chemicals were on-site. By reducing its inventory to 24 chemicals, the company saved $5,000 per year. (See page 11 for more details about Custom Print.)

For complex processes, several sets of hierarchical process maps may be needed to adequately describe all of the activities at various levels of details. Maps can include energy consumption, material flows, waste flows and air emissions. It is important to consider ancillary processes (e.g., equipment startup, maintenance), as there may be opportunities to improve these processes as well.

Figure 9 shows a sample process map for lithographic printing. The upper map provides an overview, showing only the major steps in the manufacturing process. The lower map shows greater detail of the "prepress" stage. Figure 10 then lists the inputs and wastes associated with each of these prepress steps. The focus of each map depends on the areas of interest. Figure 10, for example, could be modified to include energy consumption by various processes to allow a team to focus on identifying opportunities for increasing energy efficiency.

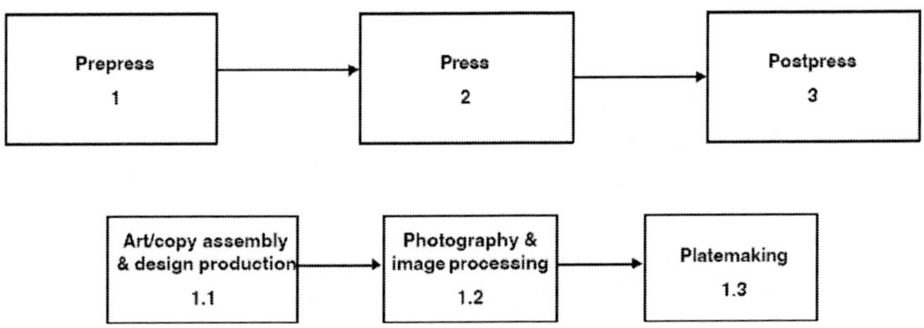

Figure 9. Lithographic Printing Process Map Example: Product Flow.

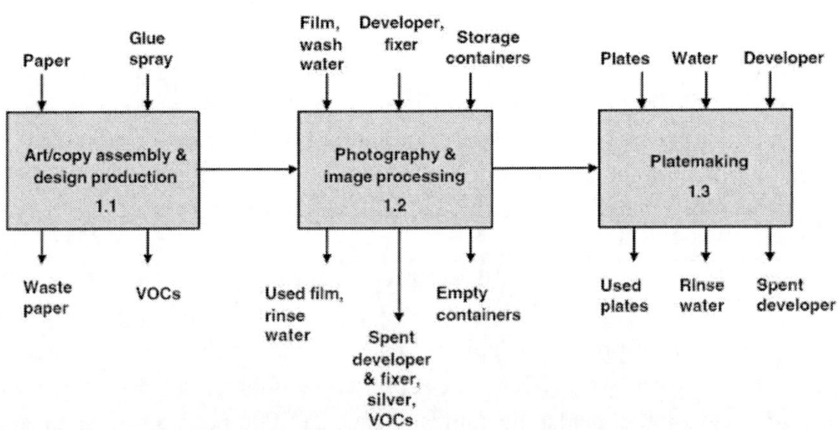

Figure 10. Lithographic Printing Process Map Detail: Input/Loss.

Resource: Systems Approach Tools

This website contains a selection of articles by Robert B. Pojasek that provide guidance on approaching process improvements from a "bigger picture" perspective.

http://www.pollutionprevention.com/Pdf-p2/handouts-PDF.html

Step 2: Identify Greener Manufacturing Opportunities

Process mapping helps you identify opportunities for improving your processes. Often elements of the process map point clearly to areas in the manufacturing process that can be improved by applying the two principles introduced earlier—(1) maximizing the percent of input material that leaves the facility as a saleable product and (2) minimizing the use of energy and ancillary materials. For example, implementing a system to reduce wasted solvents reduces the amount of emissions associated with the use of the solvent, but also reduces the need for as much solvent. Resources below can help you identify opportunities to use materials and energy more efficiently and optimize your manufacturing process to reduce waste.

Resources: On-line at Enviro-Sense

Small Business Waste Reduction Guide

37 case studies from various industry sectors (e.g., food processing, printing, wood finishing, metal cleaning, coating/painting).

http://es.epa.gov/new/business/sbdc/sbdc7.htm

VendInfo

Listings of pollution prevention equipment, products, or services (e.g., solvents, pallets, lubricants, cleaning chemicals, recovery equipment).

http://es.epa.gov/vendors/

Business Assistance Cooperative

Provides easy access to sector-based information on pollution prevention and greener manufacturing from state and local government programs, technical assistance programs, compliance assistance centers and business assistance centers.

http://es.epa.gov/cooperative/stateandlocal/

Resource: EPA Sector-based programs

Provides links to various EPA programs that focus on industry sector issues.

http://es.epa.gov/cooperative/topics/EPAintro.html

Resources: On-line Efficiency Tools

Two on-line tools are available to assist you in improving the efficiency and environmental performance of your operations.

Coatings Guide™	SAGE
The Coatings Guide™ is a pollution prevention tool for paints and coatings users. The Coatings Guide™ contains several tools to help users identify low-volatile organic compound/hazardous air pollutant coatings that may serve as drop-in replacements for existing coating operations. To date, the Coatings Guide™ information base has focused on alternative coatings for plastic and metal substrates. Watch for the addition of new information on alternative coatings for architectural substrates coming in fall 2001. http://cage.rti.org/	SAGE is a comprehensive guide designed to provide pollution prevention information on solvent and process alternatives for parts cleaning and degreasing. SAGE does not recommend any ozone depleting chemicals. http://sage.rti.org/

Resource: Technical Assistance Providers

Technical assistance providers (TAPs) help businesses identify and implement green manufacturing opportunities. You can identify a local TAP at:

http://www.epa.gov/p2/resources

or a regional TAP at:

http://www.p2rx.org

In addition to these tools, your suppliers may be a valuable information source.

Step 3: Evaluate Profitability

When capital improvements are required, it is necessary to evaluate the profitability of these greening projects. A key element in assessing the profitability of pollution prevention and greener manufacturing projects is to properly account for all environmental cost savings. Often costs (such as waste disposal) are neglected from the analysis. Techniques and software are available to help in such an evaluation (see below).

Resource: E2/FINANCE

Energy Environment Financial Analysis and Cost Evaluation System

E2/FINANCE can help you assess the profitability of potential energy efficiency or pollution prevention investments, or investments in other environmental projects.

http://www.tellus.org/risk

Resource: P2/FINANCE

Pollution Prevention Financial Analysis and Cost Evaluation System

P2/FINANCE is a spreadsheet software application for conducting financial evaluations of current and potential investments. It is unique among capital budgeting tools because it expressly addresses traditional obstacles to the financial justification of pollution prevention (P2) investments. Specifically, P2/FINANCE differs from conventional practices by expanding the cost and savings inventory to include indirect and less tangible environmental costs, and by using profitability indicators and time horizons that capture the longer-term savings typical of pollution prevention investments.

http://www.epa.gov/opptintr/acctg/download/p2finan.htm

Step 4: Strive for Continuous Improvement

Greening your manufacturing process is not something that is accomplished all at once, or something that is complete after implementation. Greening should be taken one step at a time, and always part of the decision making process.

The five stages of the "continuous improvement cycle" shown in Figure 11 are the main components of what is known as an "environmental management system" (EMS) as defined by ISO 14001 (a commonly used continuous improvement framework). An environmental management system provides a systematic way to review and improve operations for better environmental performance. It can be used to help a company improve resource utilization and efficiency, streamline operations, and meet compliance requirements.

Start Small !

Remember that you don't have to do everything at once! It is important to start small with projects that will achieve success so that you and your employees gain experience with the process and build confidence in your ability to make changes.

Even if you do not have immediate plans to embark on achieving ISO 14001 certification, the tools and techniques presented here are useful for any level of effort that seems appropriate to your company at present. The continuous improvement cycle is applicable to any business operation.

Commitment and Policy—Top management commits to environmental improvement and establishes a company environmental policy.

Planning—The company conducts a review of its operations, identifies legal requirements and environmental concerns, establishes objectives, evaluates alternatives, sets targets, and devises a plan for meeting those targets.

Implementation—The company follows through with the plan by establishing responsibilities, training, communication, documentation, operating procedures, and an emergency plan to ensure that environmental targets are met.

Evaluation—Where the company monitors its operations to evaluate whether the targets are being met, and if not, takes corrective action.

Review—The process is modified to optimize its effectiveness. The review stage creates a loop of continuous improvement for the company.

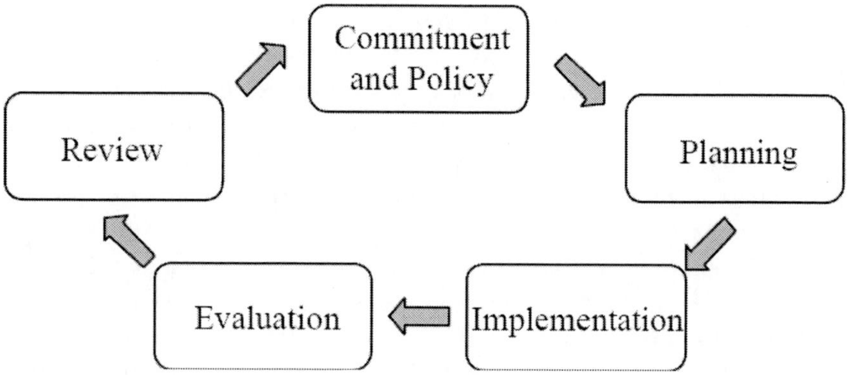

Figure 11. The Continuous Improvement Cycle.

Resources: On-Line Information About Environmental Management Systems (EMS)

Integrated Environmental Management Systems: Implementation Guide, US EPA (Oct 2000)

http://www.epa.gov/opptintr/dfe/tools/iemsguide.htm

Environmental Management Systems, NC Division of Pollution Prevention and Environmental Assistance

http://www.p2pays.org/iso/

SECTION 5. CONCLUSION

This document is intended to be an introduction to greening products, focusing on two elements of greening: (1) greening product attributes and (2) greening the manufacturing process. The following are some parting thoughts as you embark on the journey toward improving your own products.

1. **Evaluate greening opportunities during product design.** Because approximately 70% of a product's environmental impacts are determined by decisions made during its design (such as what materials are used, how the product is packaged), it is important to evaluate greening opportunities during this phase of a product's development.

2. **Greening is a moving target.** It is important to remember that the concept of "greening" is relatively new, and is therefore still a moving target. Government agencies and businesses alike are struggling to determine the best ways to improve their products to minimize impacts to human health and the environment. Replacement materials that may not be available today may be available tomorrow—always be on the lookout!
3. **Greening takes time.** It is important to remember that greening is not something that needs to happen overnight. All improvements do not have to be done at the same time. For example, with business cards, first replace the 100% virgin fiber with recycled content, increasing recycled content as material becomes available/more affordable, or moving to alternative fibers; then you can consider replacing solvent-based printing inks with soy-based inks (Figure 12).
4. **Green product market is increasing.** More and more companies, governments, and individuals are demanding and purchasing greener products. Greening your products now will give you strategic market advantages in the future.

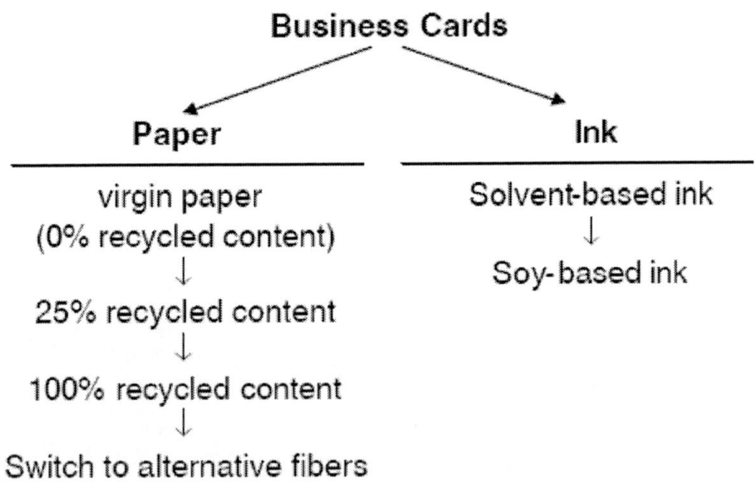

Figure 12. Example of Incremental Improvements.

Materials Inventory Sheet

Product: ____________________ Inventory conducted by: ____________________ Date: ____________________

Material/ Specifications	Quantity per unit product/ Quantity procured per month	Cost per unit product/ Cost of material purchased per month	Current vendor/ alternate vendors	Applicable alternatives/ Goals based on guidelines

NOTES:

In: Renewable Energy Products
Editor: Brett R. Harris
ISBN: 978-1-62417-565-7

Chapter 3

CONSUMER ATTITUDES ABOUT RENEWABLE ENERGY: TRENDS AND REGIONAL DIFFERENCES*

National Renewable Energy Laboratory

EXECUTIVE SUMMARY

Significant Results, Trends, and Regional Differences

The data in this report are taken from Natural Marketing Institute's (NMI) Lifestyles of Health and Sustainability (LOHAS) Consumer Trends Database® . Created in 2002, the syndicated consumer database contains responses from 2,000 to 4,000 nationally representative US adults (meaning the demographics of the sample are consistent with US Census findings) each year.

A sample of 2,000 has a confidence interval of +/-2.2 percentage points at the 95% confidence level, which decreases to +/- 1.2 percentage points with 4,000 respondents. NMI used the database to analyze consumer attitudes and behavior related to renewable energy and to update previously conducted related research. Specifically, this report will explore consumer awareness, concerns, perceived benefits, knowledge of purchase options, and usage of

* This is an edited, reformatted and augmented version of a National Renewable Energy Laboratory, Contract No. DE-AC36-08GO28308 publication, Prepared under Subcontract No. AGG-1-11880-01, dated April 2011.

renewable energy as well as provide regional comparisons and trends over time.

Based on this analysis, NMI found the following:

- The majority of consumers (80%) indicated that they care about the use of renewable energy. However, concern has diminished slightly over time, which is consistent with other broad environmental consumer attitudes (such as concerns related to environmental protection or sustainable agriculture).
- Consumers primarily associate renewable energy with environmental benefits, despite the other potential benefits renewable energy has to offer and the recent efforts to broaden its appeal.
- Consumer awareness of renewable energy purchase options remains relatively low, with approximately one in six consumers aware of the green power options provided by their electric suppliers, although approximately half of consumers have options available to them.
- Consumers are more price sensitive for renewable energy than in the past, mirroring an increased price sensitivity NMI has observed across the green consumer landscape.
- Despite the common perception, there are few differences in consumer attitudes across regions. However, consumers in the West are more aware of renewable energy terminology, such as renewable power and carbon footprint. In addition, Western consumers, compared to Midwestern consumers, are more aware of their purchase options, are less price sensitive, and are more likely to have already purchased at least some type of renewable energy.
- Seven percent of the adult population reports buying at least some renewable energy for their home. According to similar research, this is a significantly lower percentage than the proportion of the population that cares about renewable energy and a higher percentage than penetration rates reported by utilities and marketers that offer renewable energy options to consumers.

Opportunities exist for continued market growth in renewable energy as indicated by the differential between concern and usage and declining premiums.

However, the modest awareness levels of renewable energy options are a challenge.

1. Introduction

1.1. Project Background and Objective

The voluntary renewable energy market has grown considerably since its inception in the 1990s. In 2009, more than 30 million MWh of renewable energy were sold in the voluntary market, a nearly four-fold increase compared to 2005 and almost equal to the amount of new renewables required to meet 2009 state renewable portfolio standards.

Renewable energy is available to consumers in three forms: (1) utility green pricing programs offered in regulated states, (2) competitive retail electricity products in deregulated or competitive markets, and (3) renewable energy certificates (RECs) purchased in over-the-counter transactions or through independent REC marketers.

Customer participation in utility green pricing programs and competitive market programs has increased over the years, though penetration rates remain low. More than 1.4 million consumers purchase green power through their utility or retail electricity provider or in the REC market, with the vast majority (99%) of consumers purchasing through their utility or retail electricity provider. However, in 2009, the average participation rate among utility green pricing programs was 2.0%, while penetration rates in competitive markets ranged from 1.7% to 2.5%. The top 10 programs in 2009 had participation rates ranging from 5.1% to 20.8%. (Bird and Sumner 2010)

The objective of this report is to explore trends in consumer interest and awareness of renewable energy options through analysis of historical and regional survey data. Specifically, the report provides a historical and regional analysis of consumer responses to questions that indicate how much people care about renewable energy sources, how aware they are of options to buy renewable energy, their willingness to pay for renewable energy, and their opinions on the benefits of renewable energy. Only the renewable energy data are summarized here, provided by Natural Marketing Institute's (NMI's) Lifestyles of Health and Sustainability (LOHAS) Consumer Trends Database, a survey instrument conducted annually since 2002 that covers a broad range of environmental issues.

This report also provides an update to previously conducted research on consumer attitudes regarding renewable energy. Earlier consumer market research revealed consumer interest in supporting renewable energy. For example, utility market research on consumer preferences found that 52%–95% of residential consumers were willing to pay at least a modest amount

($3– $10) for renewable energy on a monthly basis (Farhar 1999). Deliberative polling of Texas electricity consumers also found that the median willingness to pay among consumers in various utility territories ranged from $1.50 to $6.50 per month (Lehr et al. 2003). In addition, a 2007 Opinion Research Corporation poll found similar results to 2007 NMI data regarding familiarity with the term renewable energy, awareness of purchasing options, and purchase rates.[1]

The updated data on consumer attitudes and awareness presented here are useful for utilities or marketers engaged in the voluntary renewable energy market. The data on trends and regional differences in attitudes towards renewable energy may also be useful for policymakers, stakeholders, and the renewable energy industry as a whole.

1.2. US LOHAS Consumer Trends Database Research Methodology

1.2.1. Survey Scope

LOHAS describes the consumers and organizations that put health and sustainability first and foremost in their lives. NMI first designed the LOHAS Consumer Trends Database in 2002 to measure and describe the marketplace for LOHAS products and services within the total US population, the consumers who use them, consumer expectations of corporate behavior, and attitudes toward environmental and social issues. Particular attention is paid to consumer attitudes, behaviors, psychographics, lifestyle activities, and product and service usage patterns in order to provide the information needed to capitalize on growing sustainability and corporate responsibility initiatives. Renewable energy has been included since the beginning of the research.

1.2.2. Data Collection

Undertaken by NMI on behalf of its sponsor clients, the 2010 research was conducted in July via an online survey of 4,000 US general population adults. Survey participants included those who had opted in to a panel as well as a demographically balanced, random subset of the panel. As respondents completed the survey, additional sample data were sent to provide balance demographics for any groups that were underrepresented at that point. To correct for any minor variations in the sample's demographic characteristics and the population as a whole, the data have been post-weighted to match multiple US Census demographic measures. No individual respondent's weighting factor is greater than three or less than 0.3. The study utilized a leading online research firm and was designed, managed, and analyzed by NMI. The text of the questions used in this report is included in Appendix A.

1.2.3. Statistical Significance

The results of this survey are nationally projectable to the US adult population and statistically valid at the 95% confidence level +/- 1.2%. Throughout this report, capital letters indicate statistically significant differences between regions at the 95% confidence level using *t*-tests. The *t*-tests reflect the probability that the difference in two numbers being compared is not due to sampling error. In regional charts and graphs, each region is assigned a letter as a short-hand code. A significant regional difference is noted on the region with the higher value by referencing the capital letter of the region with the lower value.

1.2.4. Geographic Division

This report includes analyses by geographic region. The regions are defined in Table 1 and are consistent with industry and market research standards.

Table 1. Report Categories by Geographic Region

	Northeast		Midwest		South		West
•	Connecticut	•	Illinois	•	Alabama	•	Alaska
•	Delaware	•	Indiana	•	Arkansas	•	Arizona
•	District of Columbia	•	Iowa	•	Florida	•	California
•	Maine	•	Kansas	•	Georgia	•	Colorado
•	Massachusetts	•	Michigan	•	Kentucky	•	Hawaii
•	New Hampshire	•	Minnesota	•	Louisiana	•	Idaho
•	New Jersey	•	Missouri	•	Maryland	•	Montana
•	New York	•	Nebraska	•	Mississippi	•	Nevada
•	Pennsylvania	•	North Dakota	•	North Carolina	•	New Mexico
•	Rhode Island	•	Ohio	•	Oklahoma	•	Oregon
•	Vermont	•	South Dakota	•	South Carolina	•	Utah
•	West Virginia	•	Wisconsin	•	Tennessee	•	Washington
				•	Texas	•	Wyoming
				•	Virginia		

2. Consumer Perceptions of Renewable Energy

2.1. Consumer Awareness of Renewable-energy-related Terminology, Trended

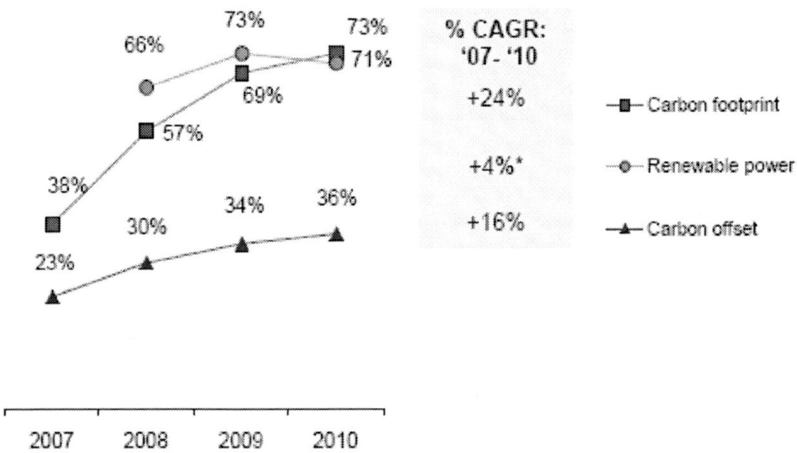

* % Compound annual growth rate (CAGR) 2008–2010[2]
Source: 2010 LOHAS Consumer Trends Database

Figure 1. In Q.14, percent of the general population indicating awareness of renewable-energy-related terminology.

Generating market interest in renewable energy starts with raising awareness of commonly used terms. Since NMI began measuring awareness of renewable power in 2008, at least two-thirds of consumers have been aware of the term, with a statistically higher number in 2010 (as shown in Figure 1). In the same timeframe, consumer awareness of carbon-related terminology has increased more rapidly. The awareness of the term carbon footprint has grown particularly quickly, up 24% annually since 2007, with most of the growth between 2007 and 2008, when media coverage of this issue was more prevalent, especially in the Northeast and West where there are established carbon regulations (i.e., RGGI in the East and Western Climate Initiative in California). As newer terms, and those commonly used by the media, many consumers have latched onto them.

The term carbon offset has not seen the same dramatic increase in awareness as carbon footprint and is still below the 2007 carbon footprint awareness level. While the two terms may be clearly linked within the business community, the same cannot be said of consumers. While carbon footprints are commonly discussed in consumer media, specifically when talking about how to reduce a carbon footprint, reports often refer to energy efficiency and conservation, use of renewable energy, and carbon offsets, among other approaches. Since carbon offsets are just one of the approaches to manage a carbon footprint, it could explain why awareness of this term is lower.

Interestingly, the term carbon footprint has overtaken the term renewable power in awareness. Since the term renewable power has been used for decades, it is remarkable that carbon footprint has eclipsed it after just a few years. While more consumers may be aware of specific types of renewable energy, such as solar and wind (85% of consumers were aware of wind energy in 2009, for instance), perhaps the umbrella term is not as well known. Therefore, communicators in this field need to be mindful of consumer familiarity with these terms in their publications.

As a comparison, 85% of consumers are aware of the term "biodegradable" (the highest scoring term measured) and 60% are aware of the term "sustainability." Awareness of both renewable power and carbon footprint, therefore, is comparable to other green industry terms. Of course, awareness is *not* the same as understanding. While awareness is a starting point, presumably fewer understand the terms. Ensuring accurate understanding is also important to generating interest in the marketplace.

2.2. Consumer Awareness of Renewable-energy-related Terminology, by Region

Table 2. In Q.14, Percent of the General Population, By Region, Indicating Awareness of Renewable-energy-related Terminology

	Northeast (A)	Midwest (B)	South (C)	West (D)
Carbon footprint	74%	71%	70%	78% BC
Renewable power	68%	71%	70%	78% ABC
Carbon offset	33%	34%	34%	44% ABC

Note: Capital letters indicate significant differences between regions at the 95% confidence level. For example, we are 95% confident that consumers in the West are more aware of "carbon footprint" than consumers in the Midwest (B) and South (C).

Source: 2010 LOHAS Consumer Trends Database.

As shown in Table 2, consumers on the West Coast are significantly more aware of these three terms than consumers in other regions who have statistically equivalent awareness. West Coast residents are often thought to be more attuned to environmental issues than consumers elsewhere in the country. However, even in the West, nearly one-quarter of the population is unaware of the term renewable power.

2.3. Consumer Caring about Using Renewable Energy, Trended

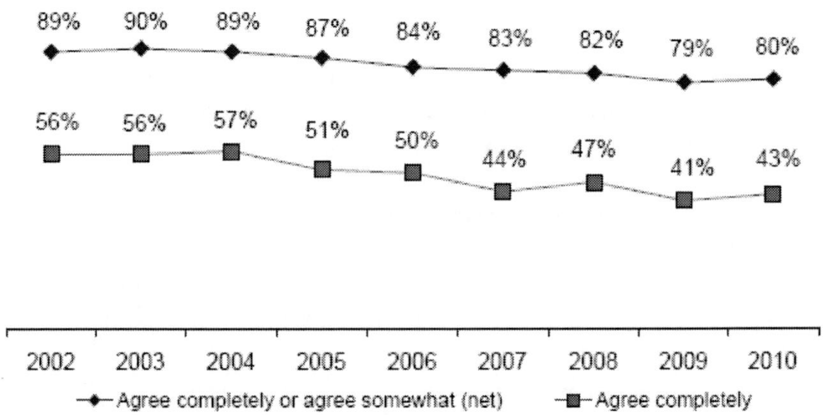

Source: 2010 LOHAS Consumer Trends Database.

Figure 2. In Q.9, percent of the general population indicating that they agree completely or agree completely/somewhat with the statement, "I care about use of renewable energy sources".

According to Figure 2, a strong majority of consumers care about renewable energy, ranging from about 80% to 90% over the survey period. Slightly fewer consumers care about using renewable energy now than they did in 2002, down 1% annually over the nine-year horizon. This decline has come primarily from the people who agree completely (down 3% annually), meaning that there is an absolute decline and a decline in intensity.

Note that more consumers report concern about use of renewable energy than are aware of the term renewable power. There are a few possible explanations for this apparent disconnect. First, the terms renewable energy and renewable power have subtle but important differences.[3] Second, the question format may affect response; for example, the awareness question asked respondents to check all they were aware of, whereas this question uses a five-point Likert scale format. A question using with a Likert scale asks respondents to answer using a scale (in this case, from "agree completely" to "disagree completely"), which is more appropriate for a question about concern or caring and also elicits a more thoughtful response.

As shown in Figure 3, concerns about other broad-based environmental issues have also experienced this decline. While the decline may seem ironic during a decade that saw significant growth in renewable energy development and considerable expansion in the number of green products on the market, the trends may actually be related.

The increased development and availability of renewable energy may indicate to consumers that the associated environmental problems have been solved, or are at least being addressed, which may allay their concerns. However, the data between 2009 and 2010 are stable, which could signal the end of the decline.

2.4. Consumer Caring about Using Renewable Energy in the Context of Other Environmental Attitudes

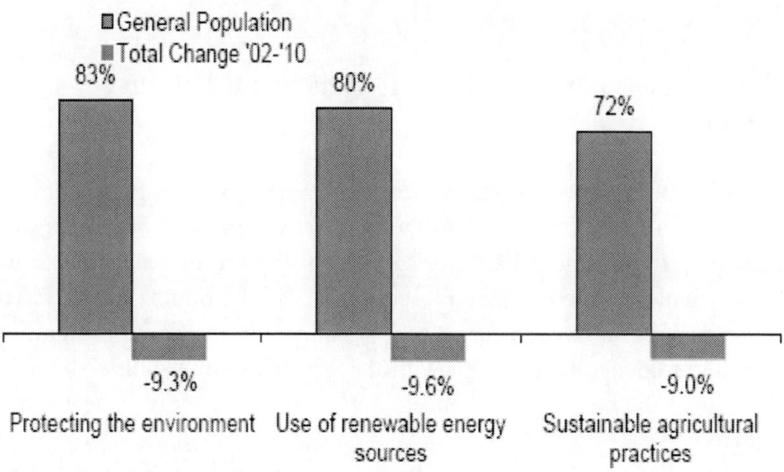

Source: 2010 LOHAS Consumer Trends Database.

Figure 3. In Q.9, percent CAGR among the general population indicating that they agree completely/somewhat with the statement, "I care about..."

As mentioned above, and displayed in Figure 3, other broad-based environmental attitudes have declined since NMI started measuring these attitudes in 2002, and the change in renewable energy is consistent with the other changes.

In addition to product proliferation potentially affecting consumer concerns, the overall level of discourse may be leading consumers to think that there is less reason for concern. Environmental issues are in the mainstream news frequently, and it would be reasonable for consumers to believe that with all of the discussion that conditions are actually improving, or are at least being addressed.

As consumer concerns appear to be diminishing, consumers are also doing more than ever to help the environment. More consumers are buying green products, for example, usage of compact fluorescent lighting (CFL) is up 15% annually since 2003 and the purchase of organic foods is up 6% annually in the same time frame. In addition, more than three times as many consumers take their own reusable bags to the store and almost twice as many say they avoid product brands with practices they do not agree with, among dozens of other behavioral changes. It is somewhat unusual for attitudes to decline while behaviors grow; however, many of these behaviors have become less burdensome for consumers, which is probably contributing to their rise.

2.5. Consumer Caring about Using Renewable Energy, by Region

As shown in Figure 4, unlike awareness of renewable energy terminology, wherein consumers in the West had the highest awareness, consumers in the Northeast are slightly more likely to care about the use of renewable energy sources, particularly in comparison to consumers in the South. The difference between consumers in the Northeast and consumers in the South is significant at the 95% confidence level. Even with the Northeastern region's slight lead over the rest of the country, the concern for renewable energy is nearly equal across regions.

This question asks consumers to pick a single, primary benefit of renewable power, though all of the choices could be seen as benefits. A stable plurality of consumers associate renewable power primarily with environmental benefits, as indicated in Figure 5.

Other benefits are much less likely to be cited as the primary benefit, even those benefits that have recently been touted heavily, namely the economic benefits.

The economic benefits are actually on a downward decline, by almost 50% from their 2007 level (a statistically significant decline). Following the trend of earlier data, few differences are evident by region, though the South is slightly more likely to associate renewable power with economic benefits than the Midwest.

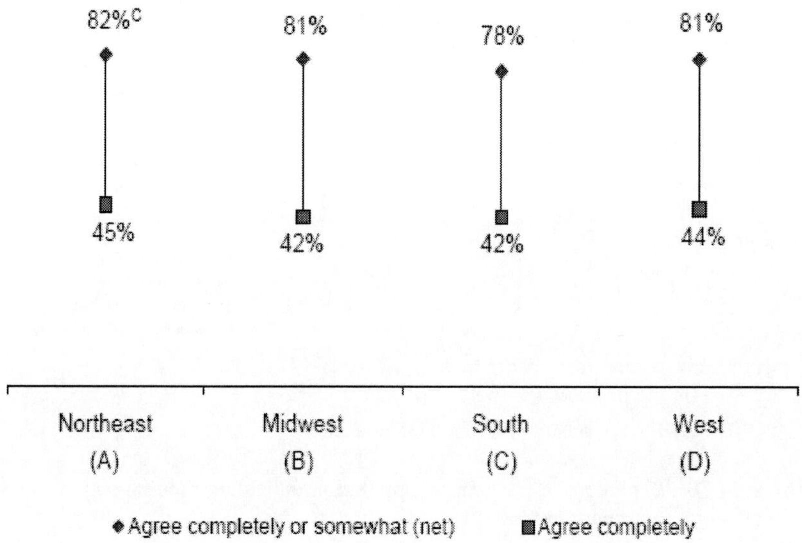

Capital letters indicate significant differences between regions at the 95% confidence level.

Source: 2010 LOHAS Consumer Trends Database.

Figure 4. In Q.9, percent of the general population, by region, indicating that they agree completely or agree completely/somewhatwith the statement, “I care about use of renewable energy sources”.

One strategy to grow renewable energy usage beyond its current level (7%) would be to better promote the health and economic benefits, since this would widen the target audience, and it is easier to reposition a product to be relevant to consumers than it is to convince them of its relevance.

As Figure 5 shows, the messaging related to health and economic benefits has not yet been effective; changing these perceptions would require very persuasive educational campaigns.

In addition, to achieve further market growth, research is needed to better understand whether these benefits are not believable, not as important, or if there are other barriers to consumer acceptance.

3. Consumer Purchasing of Renewable Energy

3.1. Consumer Awareness of Renewable Power Purchase Options, Trended

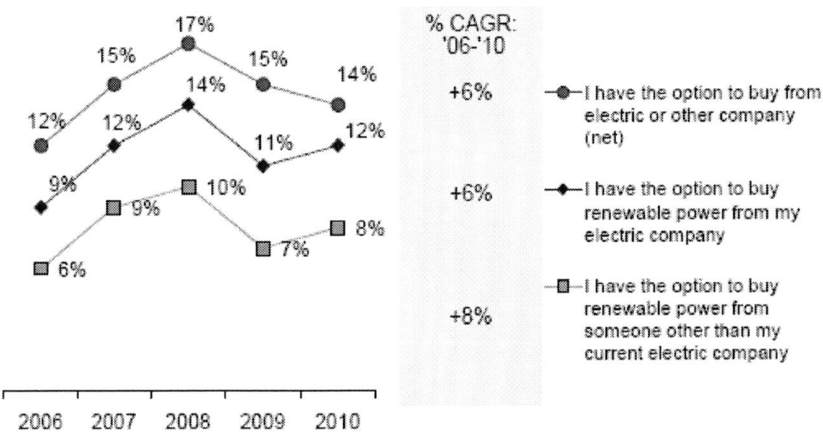

Source: 2010 LOHAS Consumer Trends Database.

Figure 6. In Q.122, percent of the general population indicating their choice of renewable power purchase options.

As depicted in Figure 6, approximately one in six adults know that they have the option to buy renewable power, either from their electric company or another provider. Consumers are much more likely to know they can get renewable power from their current electric provider since they are existing customers and receive regular communications with their monthly bills. Compared to the findings reported by NREL (Bird and Sumner 2010), the 14% of consumers who know they can buy renewable power through their utility or another company is considerably less than those who actually can (more than 50% of consumers have the option to purchase some type of green power product directly from a retail electricity provider, while all consumers have the option to buy RECs.

Compared to 2006, there has been little change in consumer awareness of purchase options, despite a peak in 2008 that did not last. Providers may have pulled back on promotions, assuming that in the height of the recession consumers had other priorities for their discretionary spending. There appears to be a bit of a correction in that, however, between 2009 and 2010.

Because electricity is such a low involvement category, meaning that consumers do not deliberate this decision extensively (if at all) or experience any consequences of their decision, these low figures are not surprising.[4] Nevertheless, there is a wide gulf between the number of consumers who reportedly care about renewable energy and those who know they can buy it (let alone those who actually do, which will be covered later in this report). If voluntary purchases are to be important to growing the renewable energy market, consumers obviously need to know they have the option of putting their money where their values are.

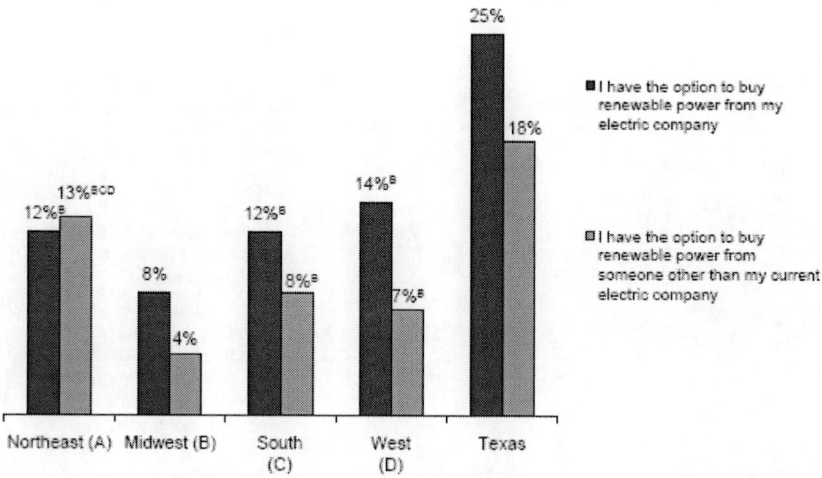

Capital letters indicate significant differences between regions at the 95% confidence level .

Source: 2010 LOHAS Consumer Trends Database.

Figure 7. In Q.122, percent of the general population, by region, indicating their renewable power purchase options.

As shown in Figure 7, there are regional variations in consumer awareness of renewable power purchase options—above and beyond most other regional differences examined in this report. Midwestern consumers are least likely to know of their options, while those on the coasts are more aware. Northeastern consumers, where retail competition exists and several alternative providers operate, are relatively more aware of both options, and the most aware of buying from an alternate supplier. In fact, this is the only region where equal numbers of consumers recognize that they have the option of buying from either type of supplier.

The most striking difference is in Texas, where one-quarter of consumers are aware of renewable power purchase options. The Texas market is one of the most competitive in the United States, with dozens of options for consumers to choose from, and Texans are most aware of their alternatives. It is also important to note that without Texas, the awareness figures for the South would be significantly lower.

3.2. Consumer Price Sensitivity for Renewable Energy, Trended

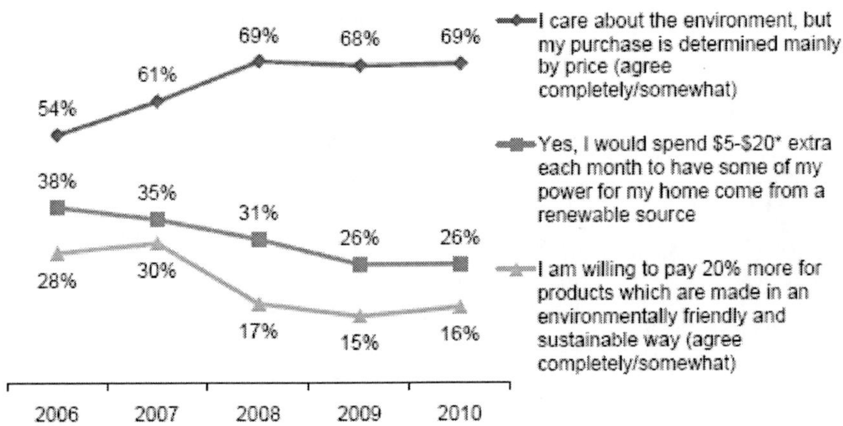

*$5–$10 prior to 2009.

Source: 2010 LOHAS Consumer Trends Database.

Figure 8. In Q.122, percent of the general population indicating their price sensitivity regarding renewable energy (trended); Q31, percent of the general population agreeing with each statement.

According to Figure 8, consumer price sensitivity for renewable energy has increased over the past five years, in conjunction with overall increased price sensitivity for other green products, as well as the overall consumer marketplace. The change between 2006 and 2010 reveals a 9% annual decline in the number of consumers willing to spend between $5 and $20 more on renewable energy for their home.

In comparison to other green products, price sensitivity is actually lower for renewable energy. In 2010, 16% of the population reports that they would pay 20% more for products made in an environmentally-friendly and sustainable way, whereas one-quarter will pay \$5–\$20 more each month for renewable energy, which covers the premium in many cases.

These figures actually far exceed the number of consumers who indicated that they currently buy renewable energy (7%).

While not everyone who says that they will pay more for renewable energy will follow through on those actions, it still suggests that there is room for market expansion.

3.3. Consumer Price Sensitivity for Renewable Energy, by Region

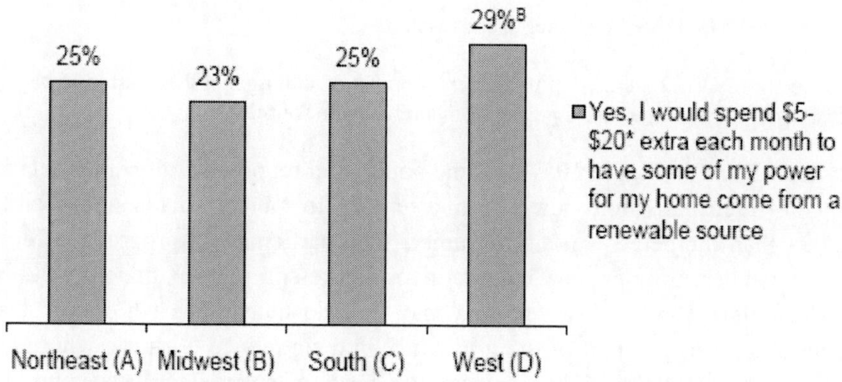

*\$5–\$10 prior to 2009.

Note: Capital letters indicate significant differences between regions at the 95% confidence level.

Source: 2010 LOHAS Consumer Trends Database.

Figure 9. In Q.122, percent of the general population, by region, stating that they would spend \$5– \$20 extra each month (\$5–\$10 prior to 2009) to have some of their household power come from a renewable source.

As shown in Figure 9, West Coast consumers are less price sensitive than those consumers in the eastern regions and statistically less price sensitive than those in the Midwest.

While these data may suggest that renewable energy marketers could charge more in the West than elsewhere in the country, it may also suggest that this region has higher market potential.

3.4. Consumer Stated Purchase of at Least Some of Their Household Power from Renewable Sources Over the Past Two Years

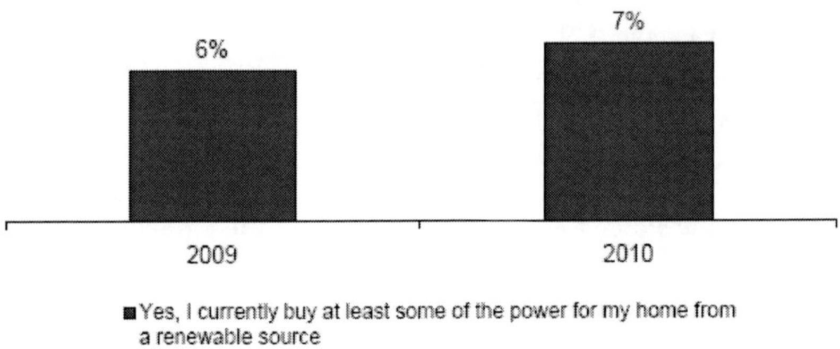

Source: 2010 LOHAS Consumer Trends Database.

Figure 10. In Q.122, percent of the general population stating that they currently buy at least some of their household power from a renewable source.

As shown in Figure 10, 7% of the population reports that they buy at least some of their household power from a renewable source. This is significantly higher than the percentage of consumers who participate in programs offered by consumer utilities and competitive marketers (approximately 1% of households). However, survey data may include individuals who know that renewable energy is part of the grid's mix in their service territory, who have solar-powered garden lights, solar water heating, solar electric systems, or other on-site renewable energy systems.

As other studies have found, this percentage is far below the portion of the population who care about renewable energy or who would pay more for it. Importantly, it is not that different from the percentage of people who know that they have the option to buy renewable energy. Simply providing consumers with greater awareness of their purchase options may be the most important factor in growing the renewable energy market.

Usage of other "green" market products may be helpful for comparative purposes (though the purchase process for these products is quite different than that for renewable energy). For example, one-quarter of the population has purchased a natural cleaning product, nearly 50% have purchased a CFL, and 30% have purchased rechargeable batteries in the past year.

As shown in Figure 11, Midwestern survey data indicates that consumers in this region are statistically less engaged than those in the West, with just 5%

stating that they buy some of their power from a renewable source. This is not surprising, since the consumers in this region are the most price sensitive and the least aware of their purchase options.

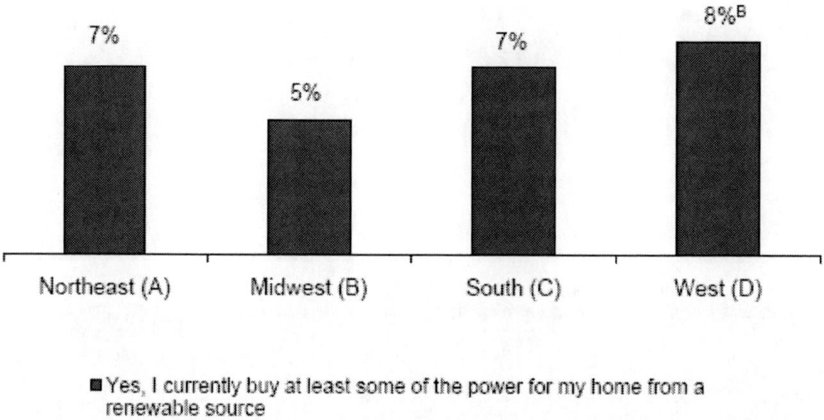

Capital letters indicate significant differences between regions at the 95% confidence level.

Source: 2010 LOHAS Consumer Trends Database.

Figure 11. In Q.122, percent of the general population, by region, stating that they currently buy at least some of their household power from a renewable source.

Conclusion

Overall, consumers have a high degree of awareness and concern for renewable energy, but few are aware of their purchase options or follow through on their stated concern, as indicated in Figure 12.

Major findings of this report include the following:

- Concern about renewable energy is quite high, with 80% of the general population caring completely or somewhat about the use of renewable energy sources. Although concern has decreased from 89% in 2002, such decline is consistent with a general decline in caring about environmental attributes, such as protecting the environment and sustainable agricultural practices.
- Consumers continue to believe that the most important benefit of renewable energy is that it is better for the environment than regular

power, although some consumers believe that the most important benefits are domestic sourcing and to a lesser extent human health or economic benefits.

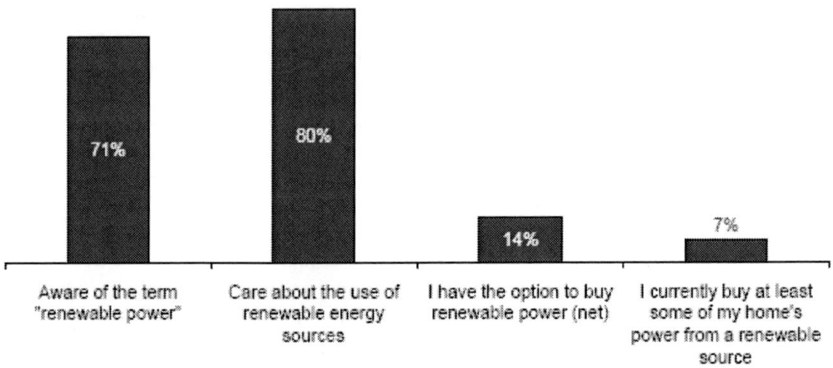

Source: 2010 LOHAS Consumer Trends Database.

Figure 12. Summary of consumer interest in renewable energy (2010 data).

- Consumer awareness is a challenge for utilities or companies providing renewable energy options to consumers. Despite a modest increase in consumer awareness, only 14% of consumers were aware of renewable energy purchase options in 2010. In addition, consumer willingness to pay more for renewable energy has declined in recent years.
- Despite the common perception that there are regional differences in attitudes about renewable energy, these data generally do not demonstrate strong regional differences, although the following differences do exist:
    - carbon footprint, renewable power, and carbon offset than consumers in other regions who have equivalent awareness. Consumers in the West are also less price sensitive than consumers in other regions and are more likely to report buying at least some of their household power from renewable sources.
    - Consumers in the Midwest are the least likely to be aware of their renewable power purchase options, while consumers in Texas and the Northeast, where retail competition exists and multiple alternative providers operate, are most aware of renewable energy purchase options.

Given low levels of product awareness, and that half of consumers who are aware of options to buy renewables do say they purchase renewables, these data suggest further opportunity for market growth in this sector, particularly if heightened consumer awareness can be achieved.

APPENDIX A. FULL SURVEY QUESTION TEXT

The questions reported in this paper appeared as follows in the 2010 LOHAS Consumer Trends Database. In some cases, other measures were included in the survey but not reported here, and those measures are not listed below.

9. Please indicate your level of agreement or disagreement with the following statements in terms of your own personal beliefs. *Please select one rating for each statement.* [*randomize*]

I care about....		Agree Completely	Agree Somewhat	Neither Agree Nor Disagree	Disagree Somewhat	Disagree Completely	Don't Understand/ Never Heard Of
a	Protecting the environment						
b	Sustainable agricultural practices						
c	Use of renewable energy sources						

14. Which of the following terms are you aware of? *Please select all that apply.* [*randomize*]

1. Biodegradable
2. Carbon footprint
3. Carbon offset
4. Sustainability

31. Thinking about *how* products are made and sold, please indicate your level of agreement or disagreement with the following statements in terms of how they affect your *purchasing* of products. *Please select one rating for each statement.* [*randomize*]

	When it comes to purchasing products or services...	Agree Completely	Agree Somewhat	Neither Agree Nor Disagree	Disagree Somewhat	Disagree Completely
a	I am willing to pay 20% more for products which are made in an environmentally friendly and sustainable way.					
b	I care about the environment, but my purchase is determined mainly by price.					

63. Which of the following *types* of products have you or someone in your household purchased in the past 12 months? *Please select all that apply.* 1. Compact fluorescent light bulbs (CFLs)

117. Please indicate the frequency with which you do the following: *Please check one for each.* [*Randomize*]

		Daily	Weekly	Monthly	Less Than Monthly	Never
a.	Not buy products from a brand or company that has practices I don't like					
b.	Take my own bag to the grocery store or other stores I shop frequently					

122. For the purposes of this survey, "renewable power" means electricity supplied from renewable energy sources, like wind and solar power. These energy sources are considered renewable sources because they are continuously replenished on the Earth. Renewable power can be used in exactly the same way as regular power. *Please select one response for each of the following.* [*randomize*]

		Yes	No	Don't Know
a	I have the option to buy renewable power from my electric/utility company.			
b	I have the option to buy renewable power from someone other than my current electric/utility company.			
c	I would spend $5–$20 extra each month to have some of the power for my home come from a renewable source.			
d	I currently buy at least some of the power for my home from a renewable source.			

123. What do you think is the most important benefit of renewable power? *Please select one*. [*randomize*]

1. Better for the environment than regular power
2. A source of energy from the United States
3. Better for human health than regular power
4. Better for the economy than regular power
5. None of the above [anchor]

122. For the purposes of this survey, "renewable power" means electricity supplied from renewable energy sources, like wind and solar power. These energy sources are considered renewable sources because they are continuously replenished on the Earth. Renewable power can be used in exactly the same way as regular power. Please select one response for each of the following. [randomize]

ACKNOWLEDGMENTS

This paper was authored by Gwynne Rogers at Natural Marketing Institute, an international strategic consulting, market research, and business development company specializing in the health, wellness, and sustainability marketplace.

This work was funded by the US Department of Energy's (DOE's) Office of Energy Efficiency and Renewable Energy (EERE). Lori Bird and Jenny Sumner of NREL served as technical monitors. The author wishes to thank Linda Silverman and the EERE technology programs for their support of this work.

The author also wishes to thank Blaine Collison of the US Environmental Protection Agency; Heidi Eggert of the Bonneville Environmental

Foundation; Bridgett Neely of Green Mountain Energy; Orrin Cook, Jennifer Martin, and Alex Pennock of the Center for Resource Solutions; and Lynn Billman, Gail Mosey, and Robin Newmark of NREL for their thoughtful review of the document, as well as Sheri Anstedt for her editorial support.

REFERENCES

Bird, L.; Sumner, J. (September 2010). *Green Power Marketing in the United State: A Status Report (2009 Data)*. NREL/TP-6A20-49403. Golden, CO: National Renewable Energy Laboratory, 69 pp.

Farhar, B. (July 1999).*Willingness to Pay for Renewable Electricity: A Review of Utility Market Research. Topical Issues Brief.* NREL/TP-550-26148. Golden, CO: National Renewable Energy Laboratory, 28 pp.

Lehr, R.L.; Guild, W.; Thomas, D.L.; Swezey, B.G. (June 2003). *Listening to Customers: How Deliberative Polling Helped Build 1,000 MW of New Renewable Energy Projects in Texas*. NREL/TP-620-33177. Golden, CO: National Renewable Energy Laboratory, 27 pp.

Natural Marketing Institute. (2002-2010). *Lifestyles of Health and Sustainability (LOHAS) Consumer Trends Database*. Harleysville, PA.

End Notes

[1] The Opinion Research Corporation 2007 poll found that 55% of consumers were either very or somewhat familiar with the term renewable energy. The poll also found that 16% of consumers were aware of a renewable energy purchase option through their utility company or electricity supplier. Of those consumers who were aware of an option to purchase, 29% reported that they did in fact purchase renewable energy through their utility company or electricity supplier.

[2] Compound annual growth rate (CAGR) is reported throughout this document. The CAGR reports growth from the first year to the most current year as if annual rates had changed at a steady rate and thus reduces the effects of year-to-year fluctuations.

[3] Renewable energy is a little broader than power, and there are forms of renewable energy that do not generate electricity, such as solar hot water.

[4] The quintessential high involvement category is a car or a house, where the purchase decision is very involved and impacts daily life. Consumer packaged goods are also a much more involved category because consumers have a more personal interaction with them (e.g., ingesting food or putting the product on the body), and the purchase cycle is rapid, meaning they are making these decisions on a weekly basis or even more often.

INDEX

D

E

F

G

H

S

T

U

V

W

RENEWABLE ENERGY: RESEARCH, DEVELOPMENT AND POLICIES

RENEWABLE ENERGY PRODUCTS

LABELING, PRODUCTION GUIDANCE AND CONSUMER ATTITUDES

RENEWABLE ENERGY: RESEARCH, DEVELOPMENT AND POLICIES

Additional books in this series can be found on Nova's website under the Series tab.

Additional E-books in this series can be found on Nova's website under the E-book tab.

ENERGY SCIENCE, ENGINEERING AND TECHNOLOGY

Additional books in this series can be found on Nova's website under the Series tab.

Additional E-books in this series can be found on Nova's website under the E-book tab.